"Jamie Zvirzdin has written a quirky, qu... lentlessly readable guide to effective science communication. Like particle physics, grammar and good writing are, at their core, a system of rules and relationships. Master these and the universe is yours!" —**Mary Roach, author of** *Fuzz* **and** *Stiff*

"Metaphor is both the food of science and the spice of language. Jamie Zvirzdin links principles of physics with those of grammar and linguistics, adds a pinch of poetics, and serves up a tasty guide to scientific thinking and writing."—**Christopher Joyce, Correspondent, Science Desk, NPR**

"By equating the fundamental particles of the Standard Model to elements of writing, Zvirzdin has created an engaging book that teaches the mechanics of science writing. Her novel approach, interspersed with fiction, nonfiction, and poetry, is both valuable and refreshing. It is a book all scientists should have on hand."—**Steven C. Martin, senior programmer at NASA / Goddard Space Flight Center**

"*Subatomic Writing*—Jamie Zvirzdin's innovative guide to telling stories of science—uses particle physics, demons, science history, family life, and a terrific sense of humor to make its points. The result is smart, effective, and a whole lot of fun."—**Deborah Blum, Pulitzer Prize-winning science writer, director of the Knight Science Journalism Program at MIT, and founder of** *Undark Magazine*

"Jamie Zvirzdin's brilliant, engaging writing guide, *Subatomic Writing*, is ostensibly here to help those who write about science, but her clear explanations, practical lessons, and gift for metaphor will help anyone struggling to assemble words into coherent strands. Who knew that the building blocks of language shared so much with particle physics? A wonderfully unique guide to vibrant writing."—**Dinty W. Moore, author of** *The Mindful Writer* **and founder of** *Brevity Magazine*

"Writing, as you may have heard, is not rocket science. In fact, it's particle physics, as Jamie Zvirzdin demonstrates in *Subatomic Writing*, her insightful new guide for the nerdy-minded scribe."—**Dava Sobel, author of *Longitude*, *Galileo's Daughter*, and, most recently, *The Glass Universe***

"Forget your preconceptions of what a book about grammar and writing is like. This is a wild ride, entertaining, enlightening, practical, and hands-on. Plus you'll learn some particle physics to boot! Zvirzdin's enthusiasm for clarity, conciseness, and vibrancy in science writing is contagious."—**Pierre Sokolsky, Dean Emeritus of the University of Utah College of Science, Distinguished Professor of Physics and Astronomy, and author of *Introduction to Ultrahigh Energy Cosmic Ray Physics***

"Inspired! *Subatomic Writing* provides an insightful and helpful new doorway into the foundations of language and writing, all wrapped up in some imaginative fun."—**Mignon Fogarty, author of the *New York Times* bestseller *Grammar Girl's Quick and Dirty Tips for Better Writing***

"This is a fantastic writing resource for people like me, people whose first language isn't English. As a guide, the book develops your writing skills, but it also tells a story, which helps you walk through the harder parts of English step by step. Easy to read, informative, and entertaining!" —**Jihee Kim, Argonne National Laboratory**

SUBATOMIC WRITING

SUBATOMIC WRITING

6
Fundamental Lessons
to Make Language Matter

JAMIE ZVIRZDIN

JOHNS HOPKINS UNIVERSITY PRESS
BALTIMORE

© 2023 Jamie Zvirzdin

All rights reserved. Published 2023

Printed in the United States of America on acid-free paper

9 8 7 6 5 4 3 2 1

Johns Hopkins University Press

2715 North Charles Street

Baltimore, Maryland 21218

www.press.jhu.edu

Library of Congress Cataloging-in-Publication Data

Names: Zvirzdin, Jamie, 1983– author.

Title: Subatomic writing : six fundamental lessons to make language matter /
   Jamie Zvirzdin.

Description: Baltimore : Johns Hopkins University Press, 2023. |
   Includes bibliographical references and index.

Identifiers: LCCN 2022018567 | ISBN 9781421446127 (paperback) |
   ISBN 9781421446134 (ebook)

Subjects: LCSH: Technical writing. | Science publishing. | Communication
   in science. | BISAC: LANGUAGE ARTS & DISCIPLINES / Writing /
   Nonfiction (incl. Memoirs) | SCIENCE / Study & Teaching

Classification: LCC T11 .Z85 2023 | DDC 808.06/66—dc23/eng/20221006

LC record available at https://lccn.loc.gov/2022018567

A catalog record for this book is available from the British Library.

Special discounts are available for bulk purchases of this book.
For more information, please contact Special Sales at specialsales@jh.edu.

To professionals and emerging professionals
who find themselves anywhere along
the Science–Language Arts spectrum:

—to scientists who want to transmit their knowledge
and experience clearly, powerfully, and memorably through writing;

—to science writers who want to rebuild
public trust in the scientific method;
and
—to creative writers, editors, and general practitioners
of the English language who want to deepen their knowledge
of the universe and share their understanding with others.

I dedicate this book to you because it is so abundantly clear to me, every time I hear of some new pseudoscience scam, medical disaster, conspiracy theory, ecological train wreck, or downright terrifying alt-reality, that we (humanity) need your science literacy, your bridge-building words, and your constructive, empathetic outreach to anchor us to what is real and good, for both the human family and our planet. Thank you for making the world a smarter, kinder space.

Contents

SUBATOMIC
WRITING

UNHALLOWED ORIGINS

Play is a strategy for learning at any age.

—Mara Krechevsky, Harvard University, Project Zero

THE NIGHT THE DEMON came to me, it was my turn to put my eleven-year-old, Maxwell, to bed. Like his nineteenth-century namesake, the physicist James Clerk Maxwell, Max is passionate about science, works extra hard in social circles, and loves playing with language. Where science and language arts overlap is a comfortable place for us, and it's roomier than people think.

Max and I were sitting on his bed that chill October night, reading Gary Larson's *The Complete Far Side*, Volume 3—yes, Volume 3—when Max turned to me.

"Why do scientists in cartoons always act like morons?" he asked.

I closed the book, stood, and pulled his Star Wars blanket over him. Our black-brown tabby cat, Tom, jumped onto the bed and curled up on Luke Skywalker's face.

"To be fair," I said, petting the cat, "everyone in cartoons, including cows and chickens, does dumb stuff. It's easy to poke fun at scientists because they're brilliant and busy, but I think people are genuinely frustrated when scientists can't—or won't—communicate information so people understand the science. Some scientists never learned how—in fact, many of my friends say they never realized they'd be writing so much in their science careers. They often say they wish they'd taken the time to learn how to write better." I scratched Tom under his chin, and the purring grew louder. "James Clerk Maxwell was brilliant in physics, but he was even more brilliant because he knew how to write about physics in a way that was accessible to others. He wasn't afraid to use analogies and play around with a little fiction to explore ideas . . . like his thought experiment with a little demon, for example."

"Maxwell's demon."

"Yes. A demon that could reverse chaos. In a letter to his friend Peter Tait, Maxwell imagined that this 'finite being' could separate hot from cold particles inside a divided box. Sorting the particles lessened the chaos in the system, but it also seemed to undermine some universal laws. Then another scientist, William Thomson—Lord Kelvin—loved the idea and called it Maxwell's 'sorting demon.' The name stuck, spread, and helped people visualize and mentally sort all those invisible particles." I bent down and kissed Max on the forehead. "You know, Gary Larson actually loves science. He wanted to study bugs as an entomologist. But his cartoons show that the curse of knowledge is a problem for both scientists and society because we don't talk to each other in a way the other can understand. If we did, I suspect we'd have fewer pseudoscience scams, more brilliant breakthroughs—and more public support for science."

Max wriggled deeper under the blanket and pulled Tom close to his chest. "You should write a book for your students," he said. He brightened. "It could be a thought experiment—how we can reverse chaos in the world through better science writing."

"Hmm, maybe." I hedged, thinking of my epic to-do list. "Anyway, g'night. Love you."

"Love you too."

I walked into the next room and turned on my computer to work on some programs for the Telescope Array Project, an international physics collaboration in Utah. Even though I earned a master's degree in writing and literature, my first love is physics, so by day, I teach science writing to graduate students at Johns Hopkins University; by night, I work on analysis programs or remotely run cosmic ray telescopes. We use these sensitive detectors in the Utah desert to track ultrahigh-energy cosmic rays, understand them, and figure out where they come from.

While Earth's atmosphere and magnetic field shield our planet from the worst of cosmic bombardment, these "rays" are actually high-speed subatomic particles, often protons, that zip through the universe and collide with our atmosphere, our Earth, and our DNA, causing small mutations within us. Outside Earth's protective magnetic field, cosmic rays can cause radiation problems for astronauts. We can't boldly go where no Shatner has gone until we figure out how to handle cosmic ray shrapnel.

While some physicists detect cosmic rays on satellites and other spacecraft, the Telescope Array Project—and its Southern Hemisphere sibling, the Pierre Auger Observatory—tracks cosmic rays from the ground. When a cosmic ray hits our atmosphere, the tiny blast creates a shower of other subatomic particles, including particles of light. The light—called *photons*, packets of energy—scatters out and hits our detectors, which we run on clear, moonless nights. With geometry and some data crunching, we see how those photons point back to where the cosmic ray entered and how much energy the original cosmic ray particle had. Where in the sky ultrahigh-energy cosmic rays come from, how massive they are, and what causes them still needs to be proven with statistical accuracy. I am fortunate to work with marvelously talented people at the University of Utah, colleagues who not only take time to answer my questions but also make the learning process enjoyable.

Although tracking subatomic particles is one of the greatest pursuits and pleasures of my life, sharing what I know through writing is another. Human communication, like particle physics, is complex, chaotic, confusing. I'm grateful my parents and English teachers read to me, listened to me read out loud, and helped me make sense of the hot mess that is the English language. Individual words shift sound, form, and meaning almost at random, jostle against other words, and combine or break down into new entities. Either by themselves or together in phrases, clauses, sentences, and paragraphs, words can be slowed or accelerated, losing or gaining momentum rhythmically or erratically.

Other words zip past our comprehension entirely, or at low energies, they rain down on our heads, impacting the surface only and never reaching our brains. A few words, however, burrow deep into us, causing critical changes to our way of thinking, our way of being. If too many words smash into us at energies too high for us to handle, we simply shut down, as some of our Telescope Array detectors do when a plane flies by on a dusty, smoky night. The words are overbright: too much information, too much noise, not enough substance. Frustrated by unexplained jargon flung at them from atop university towers, many people have likewise shut down, turning to less reliable—and more sound-bitable—sources for information.

That cold October night, while I stared at the twisted syntax of my Python program, it hit me: *particles of language are like particles of matter.* Maybe if we visualized language, especially writing, as specific sets of particles colliding and interacting with each other, we could be better science communicators and break the curse of knowledge— reverse communication chaos in a world prone to pseudoscience and error. At the very least, we could organize the chaos so readers are better able to detect and reconstruct our words with more accuracy, ease, and interest.

My mind drifted further from my program to Ada Lovelace, mathematician and daughter of the poet Lord Byron. While Charles Babbage first worked on the Analytical Engine—the humble beginnings of the mighty computer—it was Lovelace who recognized its potential. And, more importantly, she wrote about the Engine's abstract algorithm clearly and accessibly in her 1843 work, *Sketch of the Analytical Engine*, which Babbage never did. Similarly, James Clerk Maxwell pored over Henry Cavendish's chemical, electrical, and geological research and then wrote about it, distilling the brilliance of shy Cavendish and adding his own original notes. Taking time to repackage information from others in writing has itself led to innovation and insight.

My science writing students at Johns Hopkins are marvelously diverse. Some are English majors fresh out of undergrad who love science and want to include it in their fiction, nonfiction, or poetry. Some are science journalists and write articles for news outlets, magazines, or educational companies. Still others work as medical professionals and scientists—across all disciplines—and want to write grants, lab reports, white papers, articles for research journals, and nonfiction books, including memoirs. The desire to communicate is there, as is the desire to understand the universe we inhabit.

My students' knowledge of English grammar and usage is as diverse as their science backgrounds, anywhere from "I never learned this" to "I forgot everything" to "I'm already an editor." Since no writing instructor can share all they know in one semester or one book, I encourage my students to proactively fill in bits of missing knowledge by using Google, Wikipedia, *Merriam-Webster Online*, Khan Academy's free grammar course, or if they're really serious about improving their publication odds, by reading *The Chicago Manual of Style* directly, as I did in college and as my students often choose to do during the semester. There will always be some tiny issue to double-check in the style guide of your choice (or your editor's choice), but most problems I've seen over the years as a science editor point back to a few missing fundamental lessons regarding the nature of language.

Particles of language are like particles of matter. Such a fragmentary approach to writing, using a metaphor from subatomic physics, would be a risky, interdisciplinary thought experiment, like Maxwell's sorting demon: speculative rather than definitive. I thought of Michael Faraday's fear, in 1852, as he wrote about a new metaphor for magnetism—"lines of force"—in *Experimental Researches in Electricity*:

> I am now about to leave the strict line of reasoning for a
> time, and enter upon a few speculations respecting the

physical character of the lines of force. . . . It is not to be supposed for a moment that speculations of this kind are useless, or necessarily hurtful, in natural philosophy. They should ever be held as doubtful, and liable to error and to change; but they are wonderful aids in the hands of the experimentalist and mathematician. For not only are they useful in rending the vague idea more clear for the time, giving it something like a definite shape, that it may be submitted to experiment and calculation; but they lead on, by deduction and correction, to the discovery of new phaenomena, and so cause an increase and advance of real physical truth, which, unlike the hypothesis that led to it, becomes fundamental knowledge not subject to change.

Maybe physics itself could give abstract writing principles a more concrete shape. As I stared out the dark window and listened to the wind howl in the nearby woods, I thought, *We need to play with language again, bring back the spirit of speculation.* Maybe I could write a book after all. I saved the program, shut down the computer, and went to bed.

As my husband slept like the loggiest of logs on the other side of the bed—both Andrew and Max are sound sleepers, a trait I envy—I had finally drifted into dreams when something heavy sat on my chest. I couldn't move, couldn't breathe. I struggled to open my eyes. A shadow loomed over me, and the blue light from the bedside phone charger dimly illuminated the shadow's wicked features. An electric shock of terror charged through me, and then I remembered: *Sleep paralysis! Night hag! Not again. Schedule another sleep test tomorrow. Night hags are the worst!*

"Who you callin' hag, b——?" the shadow hissed.

I jerked in surprise and found I could move again. My fight response kicked in, and I kicked that hag right off the bed. Whatever it was fell with a satisfying *thud* onto the carpet.

I heard an "oof," then a groan. The shadow picked itself up. "I'm not a f——ing succubus, woman. I'm a demon. Your demon, in fact, and you just summoned me."

"What?" I whispered. "I did not."

"C'mon, 'Bring back the spirit of speculation'? Methinks thou didst, dumbass. Shall we go downstairs to talk?"

Great, I thought. *Maxwell's mother's demon.*

The demon wasn't as ugly as I'd feared, although the cheerful IKEA lighting and the sun-yellow rug in our library can make anything seem cozier, even a blue-skinned night fiend. The imp squat-jumped into one of our reading chairs, which Andrew and I had placed on either side of Tom's tall, lavish cat tree. The demon had taken my chair, of course, in the corner between the cat tree and our long row of black bookshelves. I now saw that my intruder was bigger than a baby gremlin, smaller than Chucky, better dressed than Dobby. The khaki shorts and feminine flower-print shirt, however, gave off a distinct "You interrupted my vacation" vibe.

As the demon turned and plopped down on the seat cushion, I saw her face more clearly. Her lips were coral and her hair was black and wiry, like the "mistress" in Shakespeare's famous Sonnet 130; most unnerving, however, were her three ice-blue eyes, framed by ears like a bat's, which drooped with gold earrings.

"You done ogling?" she said rudely. "I'm not anything new. Every physics student knows my dum-dum brother Theo."

"Theo? Who's Theo?"

"You call him Maxwell's demon. He's my brother. In a shocking lapse of character, Maxwell summoned *him* for help instead of me. Maxwell wanted to explore the limits of thermodynamics—that's 'heat power' in Greek for you plebeians—so Theo sorted particles between two compartments of a box. Big deal. He didn't even do any real work— just opened and closed a little trapdoor. Put fast particles here, slow particles there. Supposedly broke the second law of thermodynamics by reversing chaos. Whatever. It's not even true, by the way—Theo still spent energy keeping track of particles."

"Let me get this straight. Maxwell's thought experiment . . . *summoned* your brother?"

"Yes, of course. All thought experiments do. And like the largely useless Luke Skywalker, Maxwell's demon had an underappreciated and much more capable twin sister." She tossed back a few strands of her Crypt Keeper hair as if she were a model. "But I suppose passing gas particles has always been Theo's specialty, that flatulent douchecanoe." She tapped a long nail on her temple. "Now, trying to reverse human chaos by pushing language particles around, smashing them together, transmitting *intellectual* energy from human to human? That's a real challenge. Call it Subatomic Writing, call it Language Physics, call it Glossodynamics—Greek for 'tongue power,' haha—call it whatever you want, but it's *my* area of expertise, and it has been forever, for as long as you monkeybutts have existed. So if you're writing a book about Subatomic Writing, you're going to do it my way—the right way."

I moved from the library doorway toward the kitchen. "I need a drink," I said. "Or at least hot chocolate."

"I take single malt Scotch," she hollered. "On the rocks."

I sighed. This demon, like her brother, had clearly inherited the inability to do any actual work. When I came back, she snatched the glass tumbler from my hand with five clawed fingers, their sharp tips tinted blue like her thick skin. I remembered how Lord Kelvin, also a great writer and mathematical physicist like Maxwell, said in 1879 that he himself had found Maxwell's thought experiment "of great value" and had imagined Maxwell's demon as "endowed ideally with arms and hands and fingers—two hands and ten fingers suffice." If this demon before me was indeed related to "Theo," then at least the number of body parts was historically consistent.

"If you want to teach Subatomic Writing *correctly*," the demon said, swirling her Scotch and examining the amber cyclone with three squinting eyes, "you'll compare three aspects of writing to the three groups of subatomic particles in your current Standard Model: the *quarks*, the *leptons*, and the *bosons*. There are six of each. Six-six-six, it's easy." One of the eyes flicked from her drink to my face as if awaiting a reaction.

Hot cocoa in hand, I gingerly sank onto the couch, keeping the coffee table between us. *This is ridiculous*, I thought. Because of my sleep problems, as well as one of my favorite science books, *Hallucinations* by Oliver Sacks, I'd learned how unreliable our own normal senses are—another reason we rely on the scientific method, to triangulate reality by comparing our experiences and reliably replicating them.

But the cocoa felt hot in my hand, my gray robe was snug around me, I still heard the wind howling outside; if this was a dream, and it likely was, at least it was better than being sat on. I decided to play along. Besides, I didn't know what six bosons she was talking about. I suspected she was playing fast and loose with the traditional Standard Model of particle physics.

"OK, demon. Let's say I do write a book, a book on how to write better by thinking of language as small bits of matter. Do the six quarks represent six kinds of English words or something? Are the six leptons like . . . six general categories of typographical conventions on the page, like punctuation, italics, spaces, and such?"

"And the six bosons?" she asked quietly. Her eyes fixated again on her drink, but her body was tense, her raggedy ears alert.

"Bosons are like . . . well, they're exchange particles, transferring energy and information between other particles . . . so I suppose bosons are like the fundamental principles of great writing, the set of basic best practices to help the writer sort through and arrange all those words and typographical conventions. That way the writer's energy and information have the greatest chance of successfully transferring through the page to the reader."

"Sure," the demon said, relaxing back in my chair with a smile. It felt like I'd just passed some sort of bizarre, cosmic test. "That'll work nicely. Six kinds of English words, six kinds of typographical conventions beyond words, and six fundamental principles to help writers create combinations of language such that writers—especially science writers—actually communicate with readers. That'll do, JZ. A solid sorting. Now, before I give you my full blessing to share Subatomic Writing with the world, let's review those three piles of particles— quarks, leptons, and bosons. Do you remember the six quarks?"

"Wait. What if my students feel intimidated by the physics metaphor?" I said. "Some of them hated their physics classes."

The demon set her drink down on the nearest bookshelf. "Nonsense," she said. "They'll love it. These are the secrets of existence, the root level of all that physically exists, according to your current understanding. You people pride yourself on your roots, right? Well, this is as rooty as it gets." Hopping to the top of my chair, she used a claw to reach a high shelf and pull down one of my books. "You're the children of stars, fallen," she said, thumbing through the pages, "like Phaethon from the sun-chariot. 'Know thyself,' my boy Thales of Miletus used to say, and if thyself and thy universe are made of quarks, leptons, and bosons, then dammit, it's worth your time to learn a little bit about the itty-bitty bits that make up your much bigger biggity bits."

"But what about readers who may feel intimidated by two sets of jargon in this book, one from particle physics and one from language arts?" I persisted. "An overly elaborate metaphor just to teach writing?"

"They'll get it, bit by bit, if they take it slowly. Interdisciplinarity helps with new connections in the brain, and you'll throw in some analogies along the way to help. Make it a reference book they can come back to, with headings, lists, tables, figures, maybe a few exercises. But since you'll do your part as the writer in defining any jargon on first use, do tell them to stop being scared by big science words or gnarly grammar words. At best, they're stolen Greek words with lovely back stories, like *electricity*; at worst, they're Latin words, like *participial adjective*. You can blame the Romans for that mouthful. I do. Or, if they're not stolen words from other languages, they're words named

after people—the *proton* was named after William Prout, for example—
or they're named after a quality, like the *neutron*, which has a neutral
electrical charge. Sometimes they're even words pulled right out of
someone's . . . imagination. But we shouldn't be afraid of any words,
from anywhere. Listen to this."

She held up the book in her hands, *The Quark and the Jaguar.*
"From Murray Gell-Mann, the quark-muster himself," she said, clear-
ing her throat. She read the following passage:

> In 1963, when I assigned the name "quark" to the fundamen-
> tal constituents of the nucleon, I had the sound first, without
> the spelling, which could have been "kwork." Then, in one of
> my occasional perusals of *Finnegans Wake*, by James Joyce, I
> came across the word "quark" in the phrase "Three quarks
> for Muster Mark." . . . In any case, the number three fitted
> perfectly the way quarks occur in nature.

"See?" the demon said. "Gell-Mann assigned a nonsense word to a
real phenomenon in nature. And it's not even accurate, since quarks
don't just come in threes." She shut the book. "Know who's responsible for his 'occasional perusals' of literature? Me." She tossed my book
on the floor. "Boom. You're welcome."

"Now," she continued, "name the six quarks."

"Sorry, one more question," I said, frowning at the book on the
floor. "Do I have to mention . . . you . . . in this hypothetical book?"

"Why wouldn't you?" she said, puzzled. "Everybody loves demons—
theists, deists, atheists, orthodonists, anthropologists, stoics, hedonists,
psychics, cynics, clerics, heretics, comics, all the *-ists* and *-ics*. You monkeybutts have used us for all kinds of creative thought experiments for
a long time: from the ancient Mesopotamian *udug*, to Dante's pus-
weeping Dis, to Lucifer—Milton's all-American antihero—all the way
down to a cartoon animation of the *kikimora*, a Russian female house
spirit, on Disney's *Owl House* series."

I'd never heard of such a creature. "Are you—are you a *kikimora*?"

The demon rolled all three eyes. "I'm Greek, dummy. Your English word *demon* is connected to the ancient Greek word *daimon*, pronounced like *Simon* or *diamond*." She pointed above her, and the Greek word δαίμων burst into being as enormous flaming letters in the air. She sighed, and the letters gave out with a sparkler-like hiss. "Over time, however, my lineage was painted as more sinister than we are. I blame the Romans. Then again, I always blame the Romans."

I didn't know what she was talking about, but I resolved to look up the etymology of *demon* in the *Oxford English Dictionary*—when I woke up.

"Now, stop stalling," she said, pointing at me. "Name the six quarks."

"All right, fine. Up, Down, Charm, Strange, Top, Bottom," I recited, picking up Gell-Mann's book and returning it reverently to its place.

"Which two quarks make up 99 percent of all common matter in the universe?"

"Up and Down quarks," I replied.

"Correct. Up and Down are called *first generation* quarks, the most stable quarks in the physical universe. Similarly, in the literary universe, a writer's primary audience should understand 99 percent of all *language that matters*."

"What do you mean by 'language that matters,' exactly?"

"In your physical universe, you humans have *dark matter*—stuff you can't detect yet—and *common matter*, stuff you can detect. Ice, blood, gas, snot, dust, gold, green Jell-O—you can see this matter at the macro level, or hear it, or touch it, or taste it, or comprehend it in some way. It's sensory. With help from machines, you can extend your senses to detect matter at the micro level, the atomic level, the subatomic level. Common matter is detectable by more than just you. It exists for more than just you."

"Like how science writing should be 'detectable' or understood by the intended readers, I assume?"

"Yes. Readers can grasp this kind of writing as if it had mass. This is called, well, communication. Kind of essential for science writing."

"Subatomic Writing helps with all kinds of science writing?"

"Yes. You can create endless combinations, from Andy Weir's science fiction to Tracy K. Smith's poetry to Siddhartha Mukherjee's nonfiction to Ed Yong's journalism, for public consumption. Or more specialized science podcasts, video scripts, white papers, grants, lab memos, emails, or slide presentations to colleagues, where the primary reader already knows a certain set of science jargon, so you don't stop to define every term."

"Are you trying to put us editors out of business?"

"No, I'm doing you a favor. There are lots of different buildings in Science City, but the plumbing underneath is basically the same system. Sure, scientists can pay for editors to come in and fix poor word choices and undefined terms, unclog convoluted syntax, extract garbage, filter out unclear punctuation, repipe lengthy monotone rhythms, and reconnect logically disconnected thoughts in a paragraph. But when writers learn how to fix these flow issues themselves, it saves everyone time, confusion, and frustration."

"This is true," I said, thinking of some particularly challenging projects I'd edited, where at times the writing had been unclear to the point I was powerless to edit for clarity. When you stump even the science editor, you don't win a prize for cleverness; you need to spend more time on your writing skills. No editor, no matter how experienced, can turn garbage into gold.

"Oh, and before I forget," the demon said, "remind readers that Subatomic Writing is for second drafts and beyond. First drafts—of any kind of writing—should focus on ideas and content, not these subatomic issues. Subatomic Writing takes good drafts and makes them exponentially better."

"If I write this book, will readers need to know particle physics?" I asked.

"No," she said. "Not at all. Subatomic Writing is primarily about writing better, no more important than the zillions of other books about the craft of writing—by Joseph M. Williams, Anne E. Greene, Scott L. Montgomery, Helen Sword, Steven Pinker, Amy Einsohn, Verlyn Klinkenborg, Deborah Blum, Dinty W. Moore, on and on. Also my most favorite craft book, by Stephen King, and my least favorite, by Stunk and Blight."

I scowled. "You mean Strunk and White?"

"So damn boring. And outdated. No wonder your scientists aren't more curious about writing."

"That's not true," I protested, thinking of Max and my students.

The demon shrugged. "I speak truth to power. You've created siloed disciplines in academia, which plugs up communication channels worse than Theo plugs up the family toilet."

Seeing my disgust, the demon's jagged teeth stuck up from a thick bottom lip in what I presumed was a monstrous pout. "Tough crowd tonight, folks. All I'm saying is that good communication across arbitrary boundaries helps you all."

I softened. "That's fair."

She took another sip of Scotch and smiled nefariously. "Mm, this is good top-shelf stuff, JZ. Makes me want to come back tomorrow."

"You're coming back every night? Like Scrooge's spirits? A lesson each night until Halloween? How Dickensian of you."

"Hell, no," the demon said. "Too much work. I'm mostly here for the spirits." She raised her glass. "But since I'm here, pour me another and let's talk more about your current Standard Model. Oh, and grab a notebook and a pencil. Might as well start a first draft. Don't forget to make headings for the reader. This is for them, you know. Not you."

QUARKS:
THREE GENERATIONS OF STABILITY

Time seemed to simultaneously slow down and speed up once I refilled her glass and started taking notes. Gray graphite layers from the pencil lead streaked easily across the page as she prattled on about quarks, how Charm and Strange in the second generation and Top and Bottom in the third generation were basically bulky copies of Up and Down, but the higher generations were less stable and tended to break down—decay—back to Up and Down in the first generation.

Then, in this weird pop quiz, we reviewed how quarks could combine in twos and threes—even fours and fives—to make more complex particles. Up + Up + Down gives us our proton. Down + Down + Up gives us our neutron, a similarly stable construction.

From there, the protons and neutrons, in different numbers and combinations, create the nucleus of atoms, which then connect to other atoms, building even more complex forms of matter.

"But at the core," the demon said, "your existence as physical beings is literally dependent on the Ups and Downs." I'd never thought about life that way before, and I admit I kind of liked it.

"So," I said, trying to talk and write at the same time, "quarks clump together to create more complex, composite matter, just as words clump together to create more complex, composite thoughts for the reader."

"In general, yes. You'll explain how writers combine two or three *words* together to make a *phrase*, just like two or three quarks form protons and neutrons. Mash certain kinds of phrases together and you've got a *clause*, which is the nucleus of a sentence; likewise, mash protons and neutrons together and you've got the nucleus of an atom. Sprinkle punctuation and other typographical bits on the clause and you have a full *sentence*, just as you sprinkle in electrons and other lepton bits to create an atomic element, carbon and oxygen and whatnot. Add sentence to sentence to get a *paragraph*, add element to element in a synthesis reaction to get a compound, and boom. You're done. Then you start over again."

I considered this a moment. "You're saying writing is more about momentum than magic. Seems like more useful advice than waiting for a muse to inspire a writer with fully formed paragraphs."

"Muses suck," the demon growled, pulling another book from my shelf. "Rather than passively passing along missives from lofty sky-brats, it's way more fun to smash language together yourself. Lewis Carroll Epstein here, in *Thinking Physics*, says that physicists often focus on *mechanics*—objects moving or pushing or pulling—but tracking *collisions* is the real secret. The same goes for writing." She sat down again and began to read aloud: "Because if all the world is to be explained mechanically in terms of little balls (molecules, electrons, photons, gravitons, etc.), then the only way one ball affects another ball is if the little balls hit. If that is so, collision becomes the essence of physical interaction."

"The same is true for intellectual interaction," the demon continued, closing the book. "That is, when writers write, for any reason, they create collisions within a paragraph, within a sentence, within a clause, within a phrase, even within a word. Manage the collisions, track them, sort them, annihilate the boring and confusing particles, and writers will generate enough momentum and energy to transmit information successfully to the reader." She dumped the book on the floor.

"Stop dropping my books! They're not microphones."

"I just like the sound gravity makes," she said, crossing one leg over the other and interlacing her fingers. "Speaking of balls, if you're going to teach Subatomic Writing correctly, you will also need to talk about John Dalton's balls."

"Excuse me?"

"Are you not familiar with Dalton's balls? You know, John Dalton, father of atomic theory in England, early 1800s? Obsessive lover of lawn bowling? Took great care of his bowling balls, drew atoms as little balls for the first time, connected little wooden balls together with little wooden sticks and called it science?"

"Oh."

"Now chemistry teachers—the same ones who prohibit metaphors in science writing papers—don't bat an eye when they play with Dalton's balls. In front of students, no less!" The demon *tsk-tsked* in disapproval. "That's metaphor in action, baby, even if the metaphor is now so common people no longer see the metaphor. A ball, a line, a hole—

it doesn't have to be complicated. Or even accurate—tiny particles aren't really balls, force vectors aren't lines in space, black holes aren't holes, but all humans need a concrete metaphor as a starting point. In fact, Maxwell, when he wrote *Theory of Heat*, said that particles collide like billiard balls but not exactly, and then he explained why. So you can even acknowledge how a metaphor falls short, which leads to the reader's additional understanding of the new concept. Good science communicators connect something readers know to something they will soon know. Dalton's balls are not appreciated enough in human science history, and it's your job to fix that."

"Sure," I said, with zero intention of saying anything about poor Dalton. "Can we move on to leptons now?"

LEPTONS:
TYPOGRAPHICAL CONVENTIONS

"Fine," the demon said. She picked up Tom's brush from the cat tree and began dragging it through her own hair, mouth ajar in bliss. Seeing her beartrap-sharp teeth, I hoped my cat would stay upstairs. "Six leptons. Name them."

"The electron, muon, and tau, plus their corresponding neutrinos: electron neutrino, muon neutrino, and tau neutrino."

"Mass?" she asked. "How much matter do they have?"

"The three types of neutrinos have hardly any mass—but still some, we think. The electron is lighter than the muon, and the muon is lighter than the tau. But wait, are you not counting antiparticles in this Subatomic Writing model? Leptons each have a corresponding antiparticle, as do quarks."

She waved her free hand dismissively. "Variations on a theme. You won't be able to fit it all in one book. Antiparticles, charge, spin—maybe later. You're presenting a simplified model of both subatomic physics and language arts. You'll upset everyone from every discipline. It'll be fun."

"Great," I said. "That's exactly what I need."

She paused the hair-brushing and pulled at the eyelashes of one eye. "Eh, people will be people. You won't please everyone, no matter how well you communicate or how many drafts you write. Just be as accurate and honest as you can in the space and time you have. This is what great scientists and science writers do—or should do. Now, what does *lepton* mean in Greek?"

I massaged my forehead. I'd taught English in college and five different countries, and in the process, I'd learned a respectable level of eight other languages, including Latin, but I hadn't studied Greek at all. "I don't know," I shrugged. "All I remember is that León Rosenfeld gave those particles that name around 1950."

"*Lepton* is Greek for 'small, thin, delicate' because Rosenfeld thought leptons didn't have much mass," she said patiently. "And yet later you humans discovered that the tau lepton is wildly heavy for a particle, twice as heavy as a proton. Not small at all. This is the downside to metaphors: you need them, but once you know more, you fail to update them. Thus cosmic rays aren't rays, leptons aren't always lightweights, quantum spin doesn't spin, color charge has nothing to do with color, antimatter is still matter, and the God Particle—the Higgs boson—has nothing to do with any of the gods," she said. "Or me."

"About that," I said cautiously. "What—are you, exactly? The . . . patron saint of science writers?" I couldn't bring myself to say the word *demon* in front of her.

She threw the cat brush at my head, missing me by a fraction of a wiry hair. The brush clattered to the floor behind me. She rose to her feet—ten clawed toes digging into the fabric of my chair—and stamped her foot like a modern-day Rumpelstiltskin. "I ain't the saint of nothin', woman." She stood up straight. "I'm a proper δαίμων. I just happen to like meddling with humans in science, more frequently than my church-affiliated siblings might wish." A book fell off the shelf above me and hit my head.

"Ow!" I squawked. The book was *Screwtape Letters* by C. S. Lewis.

"See? You can't get enough of us." She picked up her tumbler again and crouched into a squat as I, tight-lipped, rubbed my head and set the book on the shelf next to me. "But I also just mess around with people. Ever heard of the Daemon of Socrates? The voice that convinced Socrates to do weird things? That was me. And his faith in positional sneezing to guide his choices? Totally me. It was hilarious."

"Wasn't Socrates killed because people considered his behavior blasphemous?"

She frowned. "A misunderstanding. I'm still upset about that."

"Can we get back to leptons?" I said. "I haven't got all night."

She swallowed an ice cube whole, not even bothering to chew it. "Yes, of course, on with the show. The six leptons in the physical universe—electron, muon, tau, and matching neutrinos—don't combine with other particles or each other the same way quarks do. Quarks connect to each another through the *strong force*, which is . . . well . . . *strong* but limited in range. An electron, on the other hand, doesn't feel the strong force at all; instead, it hangs out, statistically speaking, around the quarks, connected to them via the *electromagnetic force*. So rules for leptons are . . . looser, you could say, a bit nebulous sometimes, and heavily driven by context."

"Like how punctuation, capitalization, italics, and spacing depend on the specific sentence—as well as the writer's style and the publisher's formatting style?"

"Yes, precisely. And yet those extra writing conventions—some people call them rules, but they're just customs—are essential for better communication. That is, all those stupid commas matter more than people think, and this is why: When you write, you lose the communicative powers of physical sound and physical movement, which are three-dimensional acts that unfold during the fourth dimension, physical time. When you 'downgrade' aural speech and visual gestures to writing, you're taking dynamic four-dimensional communication to the static, two-dimensional page. The transition to timelessness comes at a cost."

Maxwell's mother's demon.

Source: Eyewitness report, as illustrated by Brent Elmer and Michael Bulla.

"So leptons in the literary universe help writers regain the power lost during that transition?"

"Not fully, but it's better than it used to be. In old-timey times, all letters of all words ran together: no punctuation, all capital letters, no word-level emphasis, no spacing. That shit's hard to read, like shouty caps on steroids. Someone—someone smart, someone Greek, of course—had to invent those extra writing bits, starting with a series of dots to mark speech pauses."

"Who?"

"Aristophanes of Byzantium, naturally. Can't believe it's already been two thousand years." She sniffed. "I miss him. Best librarian I ever had in Egypt. The Romans, of course, ignored his dots because they're freaking morons—*and* they burned down my library in Alexandria—but at least you humans had a working system by the time Gutenberg set things in type. Your English sentences now convey info more precisely with periods, commas, and the rest. The capital letters distinguish the beginning of a sentence and any other special words, making the reader notice them, and italics or boldface further emphasize what's important to the writer. Spaces clarify meaning, too. Even though they don't have much mass, like neutrinos, they still matter."

I nodded, chewing on my pencil. I was beginning to see the method in her madness.

"Yes, most madness has some method," she replied.

"Stop reading my thoughts!" I said. It was unnerving.

"Can't help it, JZ," she shrugged, setting her drink on the shelf again. "I am the product of thought experiments; therefore, I am connected to thought itself. You think, therefore I am. You're stuck with me, you lucky dog." She clambered to the top of my chair, vaulted to the highest level of the cat tree, and swung her scrawny legs over the edge. I glanced nervously toward the stairs. While Andrew and Max could sleep through the rapture, Tom was a light sleeper and rather possessive about his cat tree. I prayed—to anyone but this demon lording over my library—that Tom would not come to investigate.

BOSONS:
FUNDAMENTAL PRINCIPLES OF WRITING

"Finally," the demon said from on high, "bosons were named after Satyendra Nath Bose, an Indian physicist who worked with Einstein in the 1920s. Bosons are like a network of messengers, carrying energy and information among all other particles. More than one boson can occupy the same space at the same time, something quarks and leptons can't do. So while your literary quarks and leptons—English words and typographical conventions—usually roll out in a neat line across the page, from left to right, one after the other, your literary bosons must be a set of overlapping principles that build connections among all those letters and dots and spaces."

"Literary bosons field meaning through chaos?"

"Yes. But whatever six principles you choose must do the hard work of conducting the writer's message through increasingly complex units of language all the way to the reader."

"Look, I don't even know what six bosons you're talking about," I said crossly. "Are you counting the graviton, which is only hypothetical right now? What about the W boson? How am I supposed to know what these six lessons are?"

"I'm not gonna do all your damn work for you! Figure it out yourself," she said. She stuck her tongue out at me, a horrible forked thing with gold rings up one side of it. "Go on, I'll wait. Just think about it." She started rocking forward and backward, causing the cat tree to make a rhythmic *kathunk* noise.

I tried to ignore her as I considered what brings energy and clarity to memorable, substantive science writing. *Kathunk, kathunk, kathunk.* On a new page in my notebook, I sorted the subatomic particles and their literary equivalents into three tables (Tables 0.1, 0.2, and 0.3). The first table matched quarks to six kinds of English words, from stable to unstable. The second table matched leptons to six other bits and pieces we use in writing to supplement and clarify words. As I finished the third

TABLE 0.1

Quarks: Word families regrouped by stability, that is, readability.

		QUARKS	SYMBOL	:	WORD FAMILIES	ABBREVIATION
GENERATION	First	Up	u	:	**Common** Lexical words	lex
		Down	d	:	**Common** Function words	func
	Second	Charm	c	:	**Complex** Lexical words	comp.lex
		Strange	s	:	**Complex** Function words	comp.func
	Third	Top	t	:	**Insert** words (words that don't interact)	ins
		Bottom	b	:	**Confusing** words (outdated, slang, unnecessary jargon)	huh?

Note: Dyna told me to ignore the six antiparticles of quarks since antimatter is less common in the universe than matter. All the same, later in the text, you'll see antiquarks marked with a line above the symbol: $\bar{u}, \bar{d}, \bar{c}, \bar{s}, \bar{t}, \bar{b}$.

TABLE 0.2
Leptons: Six typographical conventions.

LEPTONS	SYMBOL	:	TYPOGRAPHICAL CONVENTIONS	ABBREVIATION
electron	e^-	:	**Punctuation** Long pause: . ? ! Medium pause: ; : — ... Short pause: , Link and Tighten: ' - .... Signal relationship to other units: ( ) [ ] – " " ' ' , , /	;)
muon	μ^-	:	**Individual emphasis** (capitalization, superscripts, subscripts)	cap
tau	τ^-	:	**Whole-word emphasis** (italics, bold, quotations, underlined)	emp
electron neutrino	ν_e	:	**Spaces between words** (one, none, or a hyphen)	sp1
muon neutrino	ν_μ	:	**Spaces between sentences** (one)	sp2
tau neutrino	ν_τ	:	**Spaces between paragraphs** (indent first line with Tab key)	sp3

Note: The six antileptons aren't shown. These conventions vary depending on purpose, audience, publisher, and the like.

TABLE 0.3

Bosons: Six fundamental principles to connect writer and reader.

CHAPTER	BOSONS	SYMBOL	FORCE OR FIELD	:	PRINCIPLE TO CONNECT WRITER AND READER	LANGUAGE LEVEL
LESSON I	**Higgs**	H	"mass force," Higgs field	:	Good Vibrations	word
LESSON II	**gluon**	g	strong nuclear force	:	Nested Classes	phrase
LESSON III	**photon**	γ	electro-magnetic force	:	Visual Syntax	clause
LESSON IV	**Z^0 boson**	Z^0	weak nuclear force, no charge	:	Mind's Breath	sentence
LESSON V	**W boson**	$W^\pm$	weak nuclear force, charged	:	Repetition, Variation	super-sentence
LESSON VI	**graviton**	G	gravitational force	:	Dot-to-Dot Game	paragraph

Note: Bosons, or exchange particles, move energy and information along; these physical forces are sometimes better pictured as fields governing the interactions and even the substance of particles. Literary bosons act as the principles that govern how energy and meaning most effectively travel through the writer's words to the intended reader.

table, I could almost see the outline of a book: each of the six lessons matched a boson to a fundamental principle of writing, giving writers the language they need to successfully transmit their message to the reader.

I held up my notebook to the demon for inspection.

She barely squinted at it over the edge of the cat tree. "Good enough," she said. "Wanna go prank your neighbors now? Got some eggs? Toilet paper?"

I was annoyed and tired. "I'm done here," I said, tossing the pencil and notebook on the coffee table and standing up. "I'm going back to bed. Do you have a name, demon?"

In a sudden burst of energy, she leapt to the coffee table—causing an inordinate amount of banging in the process. She stuck out a clawed hand.

"Name's Dyna, with a *y*. I look forward to our partnership. Don't f——it up."

I shook her hand. What else could I do? Thank goodness the hand felt like cool, cracked leather instead of tree-frog slime. Of all the bizarre things I can handle in this life with some amount of dignity, slimy demon hands are not one of them.

But it was time for this madness to end. "Nice to meet you, Dyna, but I don't want to write a book about writing fundamentals with a demon. It's been interesting, but it would be too much work, and I simply don't have the time. My students are likewise busy. No one has time for thought experiments and strange metaphors anymore. So . . . thanks, but no thanks."

I started toward the stairs. And then my poor, possessive cat wandered into the library, undoubtedly to investigate the noise. I heard a malevolent chuckle behind me. Spotting Dyna on the coffee table, Tom hissed, his back arched in full Halloween mode—and then he exploded into a million bright black-blue bits.

I whirled around. "What did you do to my cat?" I shouted.

Dyna shrugged. "He's . . . OK or not OK, depending on how well you write this week. You have six days to write a textbook on Subatomic Writing, JZ. I'll be back on the seventh night to judge the result."

I stared at the spot where Tom had vanished. A starburst-shaped burn marked the yellow rug like a sunspot. "Are you seriously Schrödingering my cat?"

Instead of answering, she turned toward my bookshelves again and pulled down Schrödinger's *What Is Life? With Mind and Matter and Autobiographical Sketches.* "Fun fact, Schrödinger didn't even have a cat in 1935, when he wrote to Einstein about how stupid quantum mechanics were. Theo's cat came back in pretty bad shape once Schrödinger was done with her. That dude was more of a prick than people realize."

"What?!"

"I said, Schrödinger didn't have a cat. He had a dog who liked biscuits. But his aunt had six Angora cats and a tomcat, and he said he hated their yowling. You didn't know about Schrödinger's aunt's cats? Don't you ever read what's in your own library?"

"I try," I muttered. "I'm busy."

"Oh, boo hoo, working mommy," Dyna said. "You must make time for things you want in life. Like your cat." She dropped Schrödinger's book roughly on the floor, and it splayed open. The sound of splattering pages made me cringe. "Now be a good girl and write the book this week, and then you'll get Tommy Boy back." She climbed onto my chair and picked up her tumbler again. I averted my gaze as she used that horrible tongue of hers to lick the final drops of Scotch from the bottom of the glass.

"One week? The whole book? Impossible!" I said, putting *What Is Life?* back on the shelf.

"Oh, come on, I gave you a major head start. An outline *and* an introduction. Because you did get one thing right, the only possible thing that could drag me away from my Caribbean booze cruise: Particles of language are indeed like particles of matter. And smashing them together is extremely fun. See you next Sunday night."

"Halloween night? Really?"

Dyna smiled, then exploded into the same million dark-bright bits as poor Tom had, leaving a similar starburst burn on my gray chair. Her tumbler fell to the floor and shattered.

✳✸

As odd as the encounter had been, I slept remarkably well the rest of the night, the best sleep I'd had in years. I opened my eyes Monday morning more cheerful than Winnie-the-Pooh in an ocean of honey. I shook off lingering whisps of the previous night with ease: the human brain is so easily duped by our own faulty senses. They say seeing is believing—maybe hearing, too, in the case of Socrates—but it shouldn't be. All the same, the dream had been refreshing, demon notwithstanding. In fact, it had almost convinced me to write a book.

"Have you seen Tom?" Max asked at the breakfast table. Andrew had already left for work. "Also, did you do a chemistry experiment last night?"

My honey-bright smile slid off my face and dripped onto the linoleum. I ran into the library. Glass shards and cat brush on the floor. Starburst burn on chair and rug. I spotted my pencil and notebook on the coffee table. Among the graphite scribbles were three tables and a doodle of a three-eyed demon.

I knew then I was destined to spend the next seven days crafting the strangest writing guide ever. I stomped back to the kitchen for a broom.

THE PARTICLE CHALLENGE

So here it goes, for poor Tom's sake, a fragmentary adventure with the particle nature of language. Since we might as well make the best of this nonsense, I invite readers to experiment with Subatomic Writing using the exercises at the end of each lesson. After you've finished the book, challenge yourself to use all six lessons to write one short, vibrant 700-word article, essay, or story teaching something about science (something accurate, please, based on reliable sources). I ask my students to use exactly 700 words, to practice concision and syntax flexibility. Then publish your writing or perform it somewhere! Feel free to join the conversation on Twitter with the tags #SubatomicWriting and @jamiezvirzdin. If, under demonic duress, you had to match up the particles of the physical and literary universes, how would you do it?

GOOD VIBRATIONS

Word Level

> The world is made of fields—substances spread through all of space that we notice through their vibrations, which appear to us as particles.

—Sean Carroll, The Particle at the End of the Universe: How the Hunt for the Higgs Boson Leads Us to the Edge of a New World

HAVE YOU EVER BEEN frustrated by vague comments from writing teachers? *Unclear*, the feedback in the margin proclaims. *Awkward. Overwritten. Jargon.* Or simply, cruelly, *WC*, which stands for either *word choice* or *water closet*, you can never be sure. But how do you fix these issues? It's easier to internalize the critique and brand yourself a poor writer than to find specific suggestions on how to improve word choice in writing.

I'm guilty of this kind of marginalia myself, in part because word choice, the first fundamental principle of Subatomic Writing, is an unstable, abstract manifestation of many other fields of study, including linguistics, history, and *semiotics*—the study of how humans create and communicate meaning. But even a brief survey of these underlying fields improves the odds that your words will resonate with readers. Per Dyna's demands, we will look at word choice anew with symbolic help from our first boson, the Higgs. This lesson might be more philosophical than the others, but we will still cover practical, concrete suggestions for improving word choice.

The Higgs boson, usually mentioned last in particle physics classes, provides the foundation for understanding physical force fields, vibrations, and particles, and it can metaphorically do the same for the literary realm. According to the Conseil Européen pour la Recherche Nucléaire (CERN, the European Council for Nuclear Research), all elementary particles were massless after the Big Bang. Soon after, the Higgs field formed. When lightning-fast quarks and leptons zipped through this force field, they gained mass—that is, they took on substance, slowed down from lightspeed. This process is called the *Higgs mechanism*, and even though the Higgs boson itself is unstable, its existence is evidence that the Higgs field bestows matter-giving properties on particles.

In science writing, we want our thoughts to pass through a literary Higgs mechanism: to slow down, resonate, and take on substance, appearing as intelligible particles of language to the reader. I think of lightning-fast thoughts zipping through the field of words stored in my brain, vibrating in my throat, slowing down and taking on substance when I talk to my students: the vibrations appear to them as words, and the words hold meaning for them (or I hope they do). Abstract thoughts thus become tangible bits of communicated information.

Even though we lose the power of audio and visual communication when we move to the flat page, as Dyna was raving about yesterday, written words can still vibrate with the sensations of sound and motion, echoes of the four-dimensional world. Which words vibrate with more life: *grow* or *proliferate*? *circle* or *circumambulatory*? *fired* or *a strategic rectification of a workforce imbalance*? No word is forbidden, and we absolutely need specialized terms, but there are multiple reasons that *grow*, *circle*, and *fired* resonate more strongly for English speakers. In a first draft, we pick whatever words we want from our mental lexicons—the dictionary we've each built up in our heads since birth. But after, when we go subatomic on the draft, we make sure the words will vibrate for the reader. At least as best as we can guess.

In the rest of this lesson, we'll sort all of English into three general word families, as linguists often do. In Subatomic Writing, these three families are further split among the six types of quarks based on how "stable" the words are for readers (see Table 0.1). We'll zero in on Up and Down words, the ones that matter most in science writing. From there, we'll examine the structure and formation of words; this background knowledge can help you decide which words will best resonate with your readers. As we map Standard English onto the Standard Model, we create a mental framework to remember how particles of language and particles of matter interact to create something bigger than the sum of their parts. I always encourage students to tinker with this baseline model: map it out on paper, reorganize it, improve it, make it work for you. A model is meant to be dynamic, not static.

The linguistic terms you'll see in this book are drawn from *The Chicago Manual of Style* and the *Longman Student Grammar of Spoken and Written English* by Biber, Conrad, and Leech. I've condensed and adapted terms to highlight science writing issues and to reduce confusion. The bold terms in this lesson are the broad divisions that map the three word families across the three generations of quarks; italicized terms fall under those broad divisions.

LINGUISTIC SORTING: THREE WORD FAMILIES

Linguists generally divide the entire English lexicon into three word families: **Lexical words**, **Function words**, and **Insert words**. I like to think of them generally as bright words, boring but necessary words, and grammatically disconnected words, respectively.

Lexical words are the main vehicles of information: these are your *nouns*, *verbs*, *adjectives*, and *adverbs*. The brighter—the more evocative—

these words are, the clearer and more interesting your message will be for the reader: *child, slither, frightened, quickly*. We'll talk more about these word classes (also called *parts of speech*) in Lesson II, but in short, they are your basic *actors, actions, details of actors*, and *details of actions*, respectively. Lexical words are the ones we see most often in newspaper headlines, such as "Blue Demon Nabs Cat Yesterday" (adjective, noun, verb, noun, adverb).

Nouns and verbs—actors and actions—are the most critical members of the Lexical word family because you'd still get the point across with the headline "Demon Nabs Cat" (noun, verb, noun). Even with complicated science topics, you can still choose strong actors who act: Particles scatter. Proteins build. Viruses die. Make your Lexical words as sensory and as concrete as you can, and keep actors close to their actions in the sentence. Great writers also choose key Lexical words (and related synonyms) to repeat, echoing the refrain-like nature of our ancestors' memorized stories. The repetition creates stronger *coherence*—connections—within and across paragraphs. More on coherence when we reach the paragraph stage (Lesson VI).

Function words, on the other hand, are purely functional. This is our second word family. Function words clarify relationships between words, phrases, clauses, sentences, and paragraphs. These are your boring words: *determiners, pronouns, primary auxiliary verbs, modal auxiliary verbs, prepositions, coordinating conjunctions*, and *subordinating conjunctions*. (These word classes are likewise described in Lesson II.) We find Function words most often before nouns and main verbs: *the, she, have, could, of, and, if*. Such old words have rattled around the English-speaking world for a long time. Great writers revise sentences so nouns and verbs—actors and actions, the story of the sentence—are not overshadowed by bossy adjectives and adverbs or too many boring Function words. This advice is as true for literary fiction as it is for any kind of nonfiction science writing.

The third word family, **Insert words**, are worth mentioning but mostly in passing. Also called Inserts, they are "stand-alone" words found mostly in speech and dialogue, usually at the beginning or end of sentences: *wow, hello, well, hey, OK?, yes, please, darn*. They don't connect grammatically with other words in the sentence, although they often need commas. Inserts include *interjections, greetings/farewells, discourse markers, attention-getters, response-getters, response forms, polite forms*, and *expletives* (see Table 1.1 for more examples). While sometimes useful in informal dialogue, you will rarely see them in formal science writing.

SUBATOMIC SORTING:
THREE GENERATIONS OF STABILITY

Our model divides the three word families among six quarks, as shown in Table 0.1 and Table 1.1. We are sorting the words now according to stability, as well as utility for the reader: What words will your specific audience comprehend most clearly? Which words are vague and destabilize your message? As Dyna mentioned yesterday, the Standard Model has three generations of quarks: Up and Down quarks belong to the first generation and possess a stable structure. Charm and Strange belong to the second generation and are less stable than the first. Finally, Top and Bottom belong to the third generation and are the least stable of the six.

Up and Down: Common Lexical and Function Words

In Subatomic Writing, we make sure that 99 percent of our draft contains **common Up and Down** words—that is, words we believe our primary audience will comprehend. They are stable: they keep your message strong and clear so the reader stays with you. Up and Down in the literary universe are your brick-and-mortar words: you lay down Lexical words and glue them together with Function words. In fact, the Greek word Aristotle used for matter, ὕλη (*hyle* or *hule*), translates as wood or timber—raw material for building.

Just as Up and Down quarks are the primary building material of all common matter, English words that readers can comprehend are

TABLE 1.1

Word families: Lexical words, Function words, and Insert words. Just as **Up** and **Down** quarks are the most common type of quark in common matter, **Lexical** and **Function** words are the most common words in writing. And just as **Top** quarks do not combine with other quarks, **Insert** words do not grammatically combine with other words in a sentence.

QUARK	SYMBOL	:	WORD FAMILY (WORD CLASSES)	WORD CLASS ABBR.
Up	u	:	**Lexical words ("brick words")** noun (*cat, boy, library, book, physics*) verb (*purr, laugh, sit, read, work on*) adjective (*soft, funny, cozy, odd, essential*) adverb (*here, today, usually, very, soon*)	n v adj adv
Down	d	:	**Function words ("mortar words")** determiner (*a, the, this, that, my, your, her, every, some, seven*, etc.) pronoun (*you, it, theirs, this, those, myself, each other, who, what*, etc.) primary auxiliary verb (*be, do, have*) modal auxiliary verb (*can, could, shall, should, will, would, may, might, must*; sometimes *ought, dare, need,* and *had better*) preposition (*in, out, on, off, of, toward*, etc.) coordinating conjunction (FANBOYS: *for, and, nor, but, or, yet, so*) subordinating conjunction (*because, while, although, if, than, that*, etc.)	det pron p.aux m.aux prep cc sc
Top	t	:	**Insert words ("stand-alone words")** interjection (*Ooh, ouch, ow, wow*) greetings/farewells (*hi, bye*) discourse marker (*well, right, now*) attention-getter (*hey, yo*) response-getter (*eh? OK?*) response form (*sure, yeah, no*) polite form (*please, thanks*) expletives (*dammit*; use with caution)	ins

Note: **Charm, Strange,** and **Bottom** (not shown) represent needlessly complicated words in science writing that should be annihilated or broken down into terms the primary audience understands. Once a term has been defined, it becomes common to the reader. The specific terms for word class will be explained in Lesson II and their abbreviations will be used in Lesson III sentence diagrams.

the primary building material in writing. If you must use a complex or unfamiliar technical word—for science writers, this is often the case—then break down the new term by providing a quick and simple definition. The unfamiliar is thus broken down to the familiar, and what is complex becomes common. If you know your audience already understands the term, then there is no need to define it: it is already a word common to both writer and reader.

Charm and Strange: Complex Lexical and Function Words

The second generation of the English lexicon also contains Lexical words and Function words, but these words are **complex** for the reader: they create a shaky foundation of understanding. These needlessly complex words are like crumbly, oversized, or oddly shaped bricks that leave cracks in your message; too little or too much mortar in between them further destabilizes meaning. Maybe you're a biostatistician writing to a biologist colleague who isn't as familiar with statistics as you are. Maybe you're a biologist writing to an NSF grant-giver who primarily studied chemistry. Maybe you're a chemist emailing a journalist. Maybe you're a journalist talking to a general audience about the biostatistician. Each field comes with its own set of jargon and stock phrases, and it's remarkably unproductive for writers to assume that readers know highly specialized terms and for readers to pretend that they understand them. This is a colossal waste of time for everyone.

We are each under the curse of knowledge, a documented cognitive bias in humans: we assume everyone knows what we know. I once went to a conference on deep machine learning where many in attendance, including me, were new to that particular field—and we struggled to keep up without first being grounded in quick, common-word definitions of extremely technical terms. Nobody there was an idiot; but some presenters forgot who their audience was, and their message failed to reach us. We are also, unfortunately, under the curse of pride: very few attendees, including me, raised their hand to ask clarifying questions, for fear of looking stupid.

A complex word in any discipline—jargon, shop talk, geekspeak—is far from evil; it just needs to be handled with care the first time it is used in a piece of writing. The **Charm** quark represents any specialized noun, verb, adjective, or adverb that the audience might not know. These **complex Lexical words** should be defined on first use and deployed without other jargon nearby when possible. Breaking down jargon by using clear definitions is akin to the second-generation Charm breaking back down to Up and Down quarks in the first generation. The more familiar you are with your audience, the better you can guess which words are common to both parties, which words need short "reminder" definitions, and which words need longer explanations.

Fortunately for writers, readers will forgive the odd word they don't know, particularly if they can pick it up from context. They may need to google a word or concept here and there. They might even understand that complex concepts take time to sink in, so if they don't understand the material perfectly before moving on, that's OK. But sometimes even colleagues need reminders as a starting point, so I suggest we err on the side of being more generous with definitions.

Some scientists and writers complain about "dumbing things down" for readers when it comes to vocabulary, and it's true that many superb words are unusable because they're unused. One of my favorite books, Ammon Shea's *Reading the* OED: *One Man, One Year, 21,730 Pages*, reacquaints us with words like *velleity* (a mere wish or desire for something without accompanying action or effort) and *bayard* (a person armed with the self-confidence of ignorance). But choosing more common words for clearer communication is the opposite of dumb. What makes us all dumb, in the end, is twisted, bloated, rushed, and self-important academese. So in Subatomic Writing, "decay" is a good thing. We take unstable Charm-like words and reduce them to stable, common, concise synonyms when possible, or we at least provide a more accessible definition on first use.

Similarly, **complex Function words** (as represented by the **Strange** quark) are boring words—often bunched together—that reduce to simpler constructions with some flexible rephrasing and substitution. For example, the phrase "despite the fact that" decays to "even though," and "regarding the matter of" decays to "about." Larger phrases and clauses can likewise decay to a stronger, simpler word. "Carefully review all that you have written thus far" can decay to one verb, "revise." There's no need to oversimplify a sentence until there's nothing substantive left, but ask yourself: Is the word count proportional to the substance? That is, do you use too many words to convey a single piece of information? Finding long strings of Strange particles and replacing them with Up and Down quarks is an easy way to improve your writing. Strange as it may seem, extricating boring words will make your writing . . . less boring. In science writing, this "strange matter" is contagious, infecting the quality of common words. Reach for more Lexical words instead—whenever possible.

What is complex versus common and boring versus bright varies from audience to audience, so make sure you identify your readership and adjust your word choice accordingly.

Top and Bottom: Insert Words and Confusing Words

Now for the third generation of literary quarks, the least useful and least stable words for science writers and their readers. We first map all Insert words to the **Top** quark. Why? Because Top quarks don't combine with other quarks, just as Insert words do not combine grammatically with the words around them. In the physical universe, Top is the most massive of the six quarks by far (heavier than most tungsten atoms!); but it is too unstable to pair up, although it can still collide and interact with particles in other ways. In the literary universe, Insert words are packed with emotion and socially rich—a great tool for dialogue—but they draw undue attention to themselves in science writing, outweighing the words around them. Again, no word is forbidden, but recognize Top's limitations.

Finally, avoid obsolete, needlessly heinous words, as represented by the **Bottom** quark. Sometimes, scientists and science writers want to sound smart, so—like quack cosmetic surgeons—they inject massive, unfamiliar words into the text to the point where readers can't even recognize the original meaning. As an extreme example, if you throw a fancy word like *floccinaucinihilipilification* at readers—and don't stop to define it as "the act of treating something as worthless"—then even your exceptionally brilliant readers will trip over it. Just as the Bottom quark eventually decays to Up or Down, you should break heavy, unstable words back down to common Lexical or Function words or, if you must resurrect some obscure words for your writing purposes, space them out and use definitions.

Bottom line: If your primary audience has zero idea what you're talking about but is impressed with how smart you are and what massive words you can use, you've failed.

And if you are defining an *obscene* number of interdisciplinary terms to the point of absurdity as part of a speculative thought experiment, throw in a demon as part of a narrative framework. If you're very, very lucky, the reader will forgive you.

IN A WORD, WORLDS: INTERNAL FEATURES OF WORDS

Ancient Greeks like Leucippus and Democritus in fifth century BCE believed the atom was the smallest possible object. The word *atom* came from *atomos*, meaning indivisible, uncuttable. Then we learned the atom *was* cuttable: into protons and neutrons and electrons, and from there into quarks, leptons, and bosons.

In 1974, Jogesh Pati and Abdus Salam cut even deeper. They suggested that quarks and leptons could be broken into point-like particles called *preons*. (Others have suggested other whimsical names.) There's no direct experimental evidence for preons—nor the one-dimensional loops of quantum string from string theory—but, who knows, maybe we'll confirm them someday.

So in the same courageous vein as Pati, Salam, and others, let's turn inward—inword—for the rest of this lesson. When closely examining any word—let's take the word *demon*, for no particular reason—we see that it has at least six internal features: **cheremes, phonemes, syllables, graphemes, morphemes**, and **lexemes**. These refer to, respectively, motion, sound, stress, written characters, meaning, and word unit. These are the individual vibrations that collectively create the written word. Each of these fundamental units—in linguistics, they're called "emic units"—is a world unto itself, but even a quick trip into each of these ideas will help you select from among the hundreds of thousands of words in the English lexicon.

1. Cheremes: Motion

In 2010, when Max was only a few months old, I took an American Sign Language (ASL) class at Gallaudet University, the only university in the world that fully operates in ASL and English. By the end of the course, I was profoundly impressed by this beautiful, complex language. There is so much we can communicate through motion alone, an ability we strive to shadow in writing.

Many linguists think of the *phoneme* as the most fundamental unit of a word (Greek, *phōnēma*, meaning "that which is sounded"). But although many ASL linguists also use the word *phoneme*, I cannot help but love the former 1960 term *chereme* (Greek, χείρ, meaning hand, plus *-eme* to match *phoneme*). A **chereme**, according to that earlier definition, is the basic unit of sign language. It is communication through motion. Cheremes in sign language are spatial and temporal, but they are just as powerful as phonemes. Maybe more.

ASL has one-handed signs, symmetric two-handed signs, and asymmetric two-handed signs. There are other parameters: handshapes, orientation of the hand, movement of the hand, location on the body or in front of it, and other gestures like moving the eyebrows, cheeks, nose, head, torso, and eyes. As for our example word, the sign for *demon* is the same sign for *devil* and *evil*: with either one hand or two, you put your thumb to your temple, index and middle finger extended as horns, and then you make "air quotes" on the side of your head, bending your horns twice.

In Subatomic Writing, we reach for this lost dimension of motion by choosing actors and actions that are graphic—that is, pictorial, vividly described, lifelike. If we must use nouns that don't use any of our physical senses, we can still reach more often for vibrant action verbs instead of the boring *be* verb, which we overuse in science writing. If you find yourself leaning too heavily on *there is . . .* , *it was . . .* , *they are . . .* , *this may be . . .* , *that may have been . . .* , *which indicates . . .* , a quick search for these verbs within your word processor (Control + F; Command + F on a Mac) will help you find and replace them with verbs that have more momentum. As often as possible, seek to increase the cheremic quality of your science prose.

The importance of motion in human communication reaches much further into the past than the earliest recorded sign-language system (which dates back to the fifth century BCE in—you guessed it—ancient Greece). In an interview with the *New York Times*, Michael Corballis, a New Zealand psychologist, said that a common ancestor could have communicated with voluntary movements of hands and face as early as five or six million years ago, with grammatical, gestural language emerging around two million years ago.

On a much smaller timescale, babies and toddlers have control of their hands and fingers before they can fully manage their vocal cords, tongues, and lips. When Max was just a few months old, he successfully communicated with Andrew and me using the two-handed sign for *more* (a Function word, used as a pronoun), his chubbly-bubbly hands forming flat *O*'s and tapping all his fingertips together. It was still a month or so before he spoke his first word: *more*. He was a hungry baby.

2. Phonemes: Sound

A **phoneme** is the smallest unit of sound in a language. Maybe it's comparable to the phonon in physics, the smallest unit of vibration. Sound itself, at its core, is not sound but motion, vibration: movement and stillness, repeated in patterns as a pressure wave that ears and brains translate into sound. If you use a lungful of air to call to someone, your vocal cords vibrate in your throat, and your mouth fashions the vibrations to form the sounds—the phonemes—into words. Phonemes can be produced all over the mouth with vibrations, hisses, and pops; vowels are shortened or lengthened through time, geography, and culture.

In human history, spoken language predated writing by many years. While a type of verbal proto-language may have begun with *Homo habilis*, about two million years ago, studies involving genetics, archaeology, and paleontology show that verbal language evolved as our vocal cords did, near sub-Saharan Africa during the Middle Stone Age, between 100,000 and 50,000 years ago—just as *Homo sapiens* was emerging. With each successive generation, tongues became more flexible, the larynx moved lower, necks became longer. Evolution favored those who successfully used motion and sound to communicate with their kinfolk.

We have around 44 phonemes in American English. Here they are, in seven groups. (Note that the line over a vowel represents what is often called the "long" sound.)

- **short vowels**: a, e, i, o, u (*cat, ket, kit, cot, cut*)
- **long vowels**: ā, ē, ī, ō, ū (*mate, meet, mite, mote, mute*)
- ***oo*-vowels**: oo, o͞o (*could, croon*)
- **diphthongs**: oi, ow (*coil, cow*)
- ***r*-controlled vowel sounds**: ar, ār, īr, or, ur (*char, chair, cheer, chore, churn*)
- **consonants**: p, b, t, d, k, g, f, v, s, z, h, j, l, m, n, r, w, y (Note that some letters of the alphabet, like *c*, are grouped under other phonemes, like *k* and *s*.)
- **digraphs**: wh, ng, ch, sh (*sure*, no vibration), zh (*measure*, with vibration), th (*thin*, no vibration), th (*this*, with vibration)

These phonemes are represented using a pair of slashes: /d/, /ē/, etc. The word *more* has only two phonemes, /m/ and /or/, a vowel whose sound is affected by the *r* that comes after it. The word *demon* has five phonemes: /d/ as in *dip*, long /ē/ as in *bee*, /m/ as in *met*, the lazy short /u/ sound *uh* as in *suppose* (often called the schwa sound), and /n/ as in *net*. Some phonemes—and their blends, like /gl/, /fr/, /sn/, and /str/—have a surprising amount of meaning already sprouting in them.

Sound Symbolism

Phonemes themselves can carry a type of primordial meaning. This fuzzy subfield of linguistics is called **sound symbolism**, although it possesses other names—phonosemantics, euphonics, or sound clusters. The imprint of meaning we find in a sound cluster amplifies the official dictionary definition (the lexeme) of a word. Sound symbolism is a subtle tool of Subatomic Writing, so faint a touch that readers probably never notice your sound choices unless you overdo it.

For the past four years, my students and I have built a—not statistically significant—list of sound clusters. It's a game of association: What is the image, emotion, or quality that bubbles up when we encounter a sound? These clusters are usually grouped around a physical object, action, or description—a common natural phenomenon we humans experience during the course of our lives. Sound clusters are grouped more by our current connotations than by the history of the word (its *etymology*), but words that come from the same root certainly show up together. As an example, words that begin with /gl/ are sometimes related to how light looks, particularly reflected light or emitted light—like quick flashes or radiating light from a burning ember. These words also describe the look of light in one's eye:

> glance, glare, glassy, glaze, gleam, glimmer, glimpse, glint,
> glisten, glitter, glitz, gloaming, gloss, glow, glower (plus
> uncommon words like gleamer, gleed, gleer, gley, glim,
> glime, glister, glogg, glore)

Not all /gl/ words conform to that proto-meaning, of course. You'll always find a word that doesn't comply: *glue, glutton, gluteus maximus.* Some of the words in our /gl/ light list stem from the Proto-Indo-European root *$\acute{g}^hley$-*, "to shine." But others, as far as etymologists can discern, were born in other old languages: Proto-Germanic, Middle High German, Old English, Middle English, Old Norse, Icelandic, and Swedish.

When we add words to sound clusters, my students and I try to avoid adding too many words with Latin and Greek prefixes and suffixes, which were adopted more directly into English from the Middle Ages onward. Many of these fractional particles (morphemes) already have a unit of meaning baked into them but can't stand on their own: the prefix *de-*, for example, signals "down, away, remove," and the prefix *dis-* is a negation of something. Sometimes we add too many prefixes and suffixes to the front and back of words, which can cause other clarity problems for readers. However, words with Latin and Greek prefixes and suffixes still wind up in sound clusters. Even though the boundaries of this game are fuzzy and more about connotation than denotation—about invoking ideas or feelings rather than a word's dictionary definition—sound clusters have the potential to convey mood, emphasize meaning, or make up for lost dimensionality in writing through the power of suggestion.

For our example word, we'll focus on the first and strongest syllable of *demon*, the /d/ and long /ē/ phonemes. Words with /d/ refer to at least eight different sound clusters, all correlated in some way to the strong, core concept of *down*.

1. **Poverty, emotional discord (to feel down):** damp, dank, dark, daunting, deadbeat, dearth, debt, deficient, dejected, depressing, derelict, deserted, desolate, despair, despondent, destitute, discouraged, disheartened, disheveled, dismal, doldrums, doleful, doom, doubt, downcast, drab, drag, dread, dreary, droop, drudge, dull, dysfunctional, shroud, sodden

2. **Delight, pleasure, tasty (to swallow down)**: debauchery, decadent, delectable, delicacy, delicious, deluxe, dessert, devour, digest, dine, divine, dreamy, drink, indulge

3. **Destroy, physical defeat (to beat down)**: damage, damn, dangerous, dashed, dazed, dead, debacle, debase, debilitate, decapitate, decay, decease, decimate, decked, defeat, defiant, defile, delete, demolish, dent, desecrate, despoil, deter, detonate, detriment, devastate, devour, diminish, disaster, discredit, dismember, dominate, downfall, draconian, drown, dupe

4. **Education, insight (to write down, get down)**: adept, adroit, address, data, debunk, define, describe, detect, dictation, didactic, diploma, discerning, discipline, discretion, doctorate, document, draw, erudition, understand

5. **Dig, little bits of ground (to go down)**: debris, deep, delve, depress, descend, detritus, dig, dip, dirt, discover, ditch, dredge, drill, dust

6. **Drops of liquid (to drip down)**: dabble, daub, dilute, distill, dot, douse, downpour, drain, dram, dreg, dribble, drip, drivel, drizzle, drop, drum

7. **Divide, sort, choose (to cut down between)**: debate, decide, deem, defect, deliberate, derail, derive, desert, detain, determine, deviate, differentiate, dilemma, direct, discriminating, dispense, dissect, distinct, divide, drive

8. **Insults (to put down)**: daffy, daft, dastard, demon, dense, deplorables, devil, ditsy, diva, drama queen, dolt, dope, dork, dotard, dotty, drivel, druggie, drunkard, duffer, dullard, dumb, dumbo, dummkopf, dummy, dump, dunce, dunderhead, and apparently "douchecanoe."

Words containing /ē/ can refer to **creepy, fearful** words:

bleed, creak, eek, eel, eerie, fearsome, fiend, fierce, flee, freaky, freeze, leech, mysterious, needle, reeling, scary, scream, screech, seedy, sheer, shriek, squeak, squeal, weak-kneed, weep, weevil, weird

But they can also refer to things that are small and potentially adorable:

baby, bumblebee, bunny, cozy, dainty, dolly, fluffy, furry, goatee, kitty, lovely, meek, peek, peep, puppy, silly, squeak, squeal, sweet, teddy, teeny, tiny, tweet, wee

My own demon, Dyna, seems to relate to most of these categories in some way.

In case you think using phonemes like this is dotty, I'm not the first science writer who encourages the use of sound to echo meaning at these small levels. Scott Montgomery, in *The Chicago Guide to Communicating Science*, writes:

Alliterative choices have real substance as well as style. They can go so far as to echo in sound what is said for meaning, or provide it with a degree of dramatic effect. *This report pays special attention to the growing mass of plastic pollution that now plagues the world's oceans.* Repeated use of *p* adds an assertive note of urgency and even accusation, like a finger tapping the table.

It helps that /p/, according to my classroom-generated sound clusters, also connotes a sense of power, action, and law enforcement (but also vanity).

His caution is the same as mine, however—to never let affectation overrule content: "Notice that even a single word change, 'growing' to 'proliferation' or 'propagating,' would again be too much and, for many readers, would even trivialize the point being made by suggesting that the writer thinks effects are more important than content."

So, feel free to use sound symbolism but, remember, it's meant to be subtle, a subatomic addition to an already good draft full of well-researched information. Use sound associations to your advantage, to help you choose words that match the mood of your topic, but not at the expense of communication.

3. Syllables: Stress

A **syllable** is a single unit of pronunciation. Syllables are weirdly more significant in the development of writing than you might appreciate. Writing in syllables developed before writing in letters, in the ancient city of Ur (present-day Iraq) in 2800 BCE.

A syllable contains one vowel sound, with or without other consonants around it. English has monosyllabic words (*cat*), disyllabic (*demon*), trisyllabic (*library*), and polysyllabic (*antidisestablishmentarianism*). The five phonemes in *demon* are spoken in two syllables: de•mon.

If a word has more than one syllable, then the word will have intrinsic **lexical stress**. The lexical stress of *demon*, for example, is on the first syllable, not the second. We say *DEE-mun*, not *duh-MOAN*. Change the stress in speech and you might lose the word. Change the movement in ASL and you might lose the word. The dictionary, besides showing how to spell, pronounce, and divide a word into syllables, will also show primary lexical stress (and secondary lexical stress, if there is one). In *demon*, therefore, the **stress** is on the first syllable, and the second syllable is **unstressed**.

Almost all simple signs are monosyllabic in ASL. Similarly, some of our best common words are one syllable long. I recommend skimming the Wikipedia entry "Most Common Words in English." These words are short, fast, communicative. They often have enormous entries in the dictionary, yet we're still able to process them more quickly than complex words with longer syllables and more exact meanings.

Since syllables also strongly influence the flow of a sentence, we will return to syllables—strung together in a sentence, creating **prosodic** rhythm—in Lesson V.

4. Graphemes: Written Characters

Graphemes, finally, are our alphabet characters, our fundamental writing tool kit. We draw graphemes on a page—or, with a sparkler on the Fourth of July, in the air—from left to right. Our example word, *demon*, represents itself here with the letters *d, e, m, o, n*—a series of lines and curves, nothing more than wiggly little characters strung across space.

With 44 phonemes but only 26 letters, we must get creative with graphemes to catch all the phonemes of English—hence the digraphs and diphthongs and whatnot. This is part of the reason English spelling is all over the place (the other reason is historical, as described in the upcoming section on morphemes).

The first writing systems of humankind include the Jiahu symbols in China during the Neolithic period—around seventh millennium BCE. A series of cuneiform marks in ancient Mesopotamia is another, about 5500 years ago. Writing is still a relatively recent invention, and it hasn't finished evolving. Latin rules, forced onto English writing in the Middle Ages, did not reflect how we were speaking for hundreds of years; this grip of Latin rules, in *Chicago* style and elsewhere, has now loosened on split infinitives and on ending sentences with prepositions. Slowly, we grow and change and improve.

Sound and Motion, Reborn

The vibrations within words tangle even more now: believe it or not, our brains still process graphemes as if they were phonemes. Even though your audience may be reading silently, skimming eyes over letters, their brains still hear and respond to the sound of your writing. If you need more proof, I invite you to look up the *Scientific American* article "When We Read, We Recognize Words as Pictures and Hear Them Spoken Aloud" by Stephani Sutherland, and *Discover Magazine*'s "In the Brain, Silent Reading Is the Same as Talking to Yourself" by Carl Engelking. From what researchers can tell, we may see the shape of the word, but we still hear the word in our minds.

Consider this illuminating letter snippet from Isaac Asimov to Carl Sagan in 1973: "I have just finished *The Cosmic Connection* and loved every word of it. You are my idea of a good writer because you have an unmannered style, and when I read what you write, I hear you talking." To me, this exchange between two great science communicators means two things: one, that the sound of a word—its phonemes—should continue to be a factor as we choose our words; and second, you can write with common ("unmannered") words readers understand and still gain admiration from other great writers.

Not to be outdone, motion and shape—cheremic qualities—make a remarkable appearance even at the level of graphemes. In Figure 1.1, which shape is named *bouba* and which is named *kiki*?

FIGURE 1.1 The bouba/kiki effect, an example of sound symbolism. When people connect sounds and shapes, the connections are not always arbitrary. Most people assign the shape on the left *kiki* and the shape on the right *bouba*.
Source: Jamie Zvirzdin.

The "bouba/kiki effect" happens when we associate the "voiced" (vibrational) sound of the nonsense word *bouba* with a round shape; similarly, we match the "unvoiced" sound of the nonsense word *kiki* to a spiky shape. While this study originally centered on the connection of sound and an abstract shape—and on the motion of the words in our mouths—two 2017 studies by Christine Cuskley, Julia Simner, and Simon Kirby suggest that what heavily influences a reader's decision is the shapes of graphemes and the abstract shapes. The word *bouba* itself is a rounded set of shapes. *Kiki* itself is a "spiky" word. The researchers showed that for literate subjects, the curvature of letters is strong enough to significantly influence word–shape associations, even in auditory tasks.

Even though the evolution of each of our alphabet letters—our primary graphemes—is quite messy, the Minds and Traditions Research Group (the Mint), led by Olivier Morin at the Max Planck Institute for the Science of Human History, seeks to renew the study of nonverbal graphic codes and how they are transmitted over time by cultures. As part of this goal, they study what makes shapes easier or harder for our eyes to process:

> Complex shapes, with a high ratio of contour to surface, take more time to process and tend to have less appeal than simple ones. (Letters like O or C are simpler than W or Q in this sense.) Cardinal straight lines (horizontals and verticals, as in H or E) are processed more fluently than oblique ones (as in W or X). The scripts used by the writing systems of the world must weigh these demands of visual simplicity against the need to encode large quantities of information (which overly simple or uniform shapes cannot do). Finding an optimal balance between informativeness and simplicity is a difficult task.

As you find your own fulcrum for this balancing act in science writing, remember how sound, motion, and shape affect the reader's experience of meaning. This phenomenon, also known as iconicity, can help you choose how to represent multidimensionality on the page.

5. Morphemes: Meaning

Morphemes are the smallest unit of meaning in a language. They can be **free morphemes**, able to stand on their own, or they can be meaningful fragments of a word—**bound morphemes**—added to the **root** unit of meaning.

For example, *demon* is no longer a proper *demon* if you drop a letter. If we cut up the word in various places, we get *dem, demo, em,* and *emo.* These are all free morphemes with separate meanings from *demon.* Cut the word differently and you also get the prefixes *de-* and *mon-*, as well as the suffix *-on,* which we see a lot in Subatomic Writing. (The suffix *-on* is added to the end of words to form nouns that denote particles, molecular units, or substances, such as *proton, neutron, electron.*) However, those prefixes and suffixes are bound morphemes; they must be mashed onto another word to make their meanings collide and fuse. Therefore *demons,* plural, has two morphemes, a root morpheme plus a bound morpheme: *demon + -s.*

We are lucky to have a rich field of words in the English language. Each morpheme has a history—some fascinating, some boring. With a little regard for the etymology of a morpheme, writers can choose morphemes that carry meaning more substantively and efficiently to the reader.

Digging deep into the field of words, English is, at its core, a Germanic language, not French, not Latin, and not Greek. As much as Dyna would like to Make America Greek Again (too soon?), only about 6 percent of our words were adopted from Greek. Here's a simplified summary of what happened, according to Kenneth Katzner's *The Languages of the World.* Around the fifth century CE, from a small "angle" or corner of land in present-day Schleswig-Holstein, Germany, the Angles tribe arrived in Britain. They, along with two other Germanic tribes, the Jutes and Saxons, invaded Celtic land. The Germanic dialects, written in runes, ruled the linguistic landscape. These tribes were in turn invaded by Vikings, which added Old Norse to the language. Old English and Old Norse competed for supremacy and, while one or the other won out for a specific word, we often kept both and created nuanced distinctions between the two terms. Consider these Norse/English pairs: *ill/sick, bask/bath, skill/craft, skin/hide, dike/ditch,* or the Norse-versus-English pronunciation of words (*sk* versus *sh*) that split and took on different meanings over time: *skirt/shirt, scatter/shatter, skip/shift.*

Then Christian missionaries from Ireland and Rome turned the runes into the Roman alphabet we know, and they added a layer of Latin to the language. The Norman Conquest of 1066 added a French layer to this already impressive cake of language. Whereas English farmers said *cow, calf, ox, deer, sheep, swine,* the French aristocracy, mouths full, said *beef, veal, mutton, venison, pork, bacon.* There's a word for everything, plus a host of synonyms.

English has only grown and added more morphemes since then, particularly during science, technology, religious, and cultural revolutions. Despite the many layers, the Old English of Anglo-Saxon times vibrates at the core of modern English: in everyday household words and in most words for body parts, plus pronouns, prepositions, conjunctions, and auxiliary verbs (Function words!) at the heart of our grammar.

Whenever you can, choose Germanic morphemes over Latin and Greek ones. Why? They are old, short, and powerful. They are the heartstrings of English. Pluck them, make them vibrate. If you don't have time to learn Latin, Greek, German, etc., you can still learn which words came from where by bookmarking a very nice, epic entry on Wikipedia called "List of Germanic and Latinate Equivalents in English." When revising, use this list to switch at least some Latinate words to Germanic if you can. If you're still struggling, see if you can identify the root morpheme of your Latin or Greek word using the Wikipedia entry "List of Greek and Latin roots in English." What is the equivalent word in the "Meaning in English" column? Whenever possible—and it's not always possible or advisable—translate your Latin and Greek words to the deeper Germanic root of the English lexicon. These are the common words readers digest most quickly and easily.

I absolutely understand that in science writing, "formal" has come to mean "filled with Latin and Greek words." But formal does not necessarily mean accessible or even intelligible, and the formality alienates people who lack the time, privilege, or patience to learn the extra layers of English. Weirdly, this linguistic alienation is still felt by people who have had the time, privilege, and patience to learn the extra layers. I loved learning Latin in college; I learned it so well I tutored other students as a part-time job. I loved learning French and speaking it when I lived in Belgium. I loved talking with classmates and friends who were studying ancient Greek and ancient Egyptian. And yet, when I'm reading about science in English, I will take *grow* over *proliferate* any day of the week.

You'll have pushback from people in science, including science editors, who want to preserve formality in the name of authority. I get it. But as Martin Luther discovered when he translated the Bible into "vulgar" German, and as William Tyndale discovered when he translated the Bible into "vulgar" English, there is immense power in choosing words attached to the heartstrings of your home language. You'll discover ways to preserve a formal tone even if you use common words.

Dezombifying

A major problem science writers must face, even if they do choose words with Germanic roots, is their addiction to morphemes. We must wean ourselves off them. Words with a glut of morphemes packed onto the root morpheme are **nominalizations**. "Zombie nouns," as Helen Sword calls them. Take adjectives, verbs, or nouns and add suffixes to them like -*ity*, -*tion*, or -*ism*, she says in her 2012 *New York Times* article. You've just created new, horrible nouns:

> The *proliferation* of *nominalizations* in a discursive *formation* may be an *indication* of a *tendency* toward *pomposity* and *abstraction*.

In attempting to sound academic and intelligent, you've created a sentence, Sword says, where zombie nouns "cannibalize active verbs, suck the lifeblood from adjectives and substitute abstract entities for human beings." We amp up clarity and interest when we go back to the original (root) morpheme:

> Writers who overload their sentences with *nominalizations* tend to sound pompous and abstract.

We can use nominalizations if we absolutely need them—sometimes we need them—but otherwise, go for the root. Long strings of long words do not signify we are smart; it means we've forgotten how to communicate effectively.

As Winston Churchill said, "Short words are best, and old words, when short, are best of all."

6. Lexemes: Word Unit

Lexemes are what we often think of when we say "words." A lexeme is the basic abstract unit of meaning of a single word. Similar lexemes (*talk, talks, talking*) are grouped together in the **headword** of the dictionary definition. From the shelf in my office, I hefted down Volume D of my 1978 *Oxford English Dictionary* (*OED*), which I found buried deep in an enormous secondhand bookstore. Although I use *Merriam-Webster Online* on a regular basis when I edit, I like to use the *OED* when I write because it records a word's origins, at least as far as word-archaeologists (etymologists and lexicographers) can recover from the field of language.

Deadwood . . . deconstruct . . . demon. There it was. Next to the headword *demon* was its pronunciation, dī mən (the upside down *e* represents the schwa sound), in two syllables, along with an alternate spelling—*dæmon*, pronounced the same way. The first definition confirmed Dyna's words: "In ancient Greek mythology (=δαίμων): A supernatural being of a nature intermediate between that of gods and men; an inferior divinity, spirit, genius." It was followed by an unnerving quotation showing *dæmon*'s earliest usage in English:

> Grammarians . . . doo expounde this word Dæmon, that is a
> Spirite, as if it were Sapiens, that is, Wise.
>
> —Henricus Cornelius Agrippa, translated from Latin to English
> in 1569, *Of the Vanitie and Uncertaintie of Artes and Sciences*

A quick check of the *OED Online* showed that the ancient Greek meaning had slipped from the first definition to the eighth since 1978. The first definition is currently: "Any evil spirit or malevolent supernatural being; a devil." In between the first and the eighth definitions ranged a whole host of meanings: an idol, a mischievous child, an ugly animal, alcohol, recurrent fears or compulsions, a strong or skillful person, an enthusiast, an Australian police officer (slang), an escaped convict (rare), a British card game (huh?), and a person's spirit animal, referencing Philip Pullman's *His Dark Materials* trilogy. The tenth definition referenced Maxwell's demon, but also Laplace's demon and Szilard's demon, with the definition, "Any of the various notional entities having special abilities, used in scientific thought experiments." Maxwell's mother's demon—with the definition, "An annoying, didactic trickster and cat-nabber promoting Subatomic Writing specifically and the merging of the arts and sciences generally"—was nowhere to be found.

Looking up *dæmon* directly online revealed a computing term, a computer program that runs in the background, like the "mailer daemon" that sends you an automatic message if your email bounced. Maybe another daemonic relative was responsible for all bounced emails.

Digging further, by this point feeling like a character in a Dan Brown novel, I saw an alternate spelling and pronunciation, *daimōn*, like *diamond*, and it had additional spinoff meanings to the Greek: a supernatural being of inspiring force, a tutelary deity. At the core seemed to be the Greek word *daiomai*, meaning *to divide, lacerate, assign*. A *daimōn*, therefore, meant "the lacerator, one who divides and fractures." Sorting demon indeed.

If I was going to get my cat back, however, I'd better pull out of Illuminati mode and finish this lesson. Before I shut the dictionary, however, I also noted how the word *demon* can be **inflected**, **derived**, or **compounded** by adding morphemes to the base word.

Inflections in English don't change the base lexeme of a word, but they do alter the word to signal small variations within the base meaning. The noun *demon*, for example, has three inflections. In the first inflection, the base word can take on the morpheme *-s* and become *demons*, meaning more than one demon. If I added the morpheme *-'s*, with an apostrophe, I'd get the *possessive* form, meaning that the demon owns or possesses whatever noun comes next: *the demon's victim*. You can unpack the possessive in English by kicking *demon* to the other side of the noun and adding *of*: *the victim of the demon*. This is useful later if you're having trouble with adjective or preposition strings and need to reword a phrase.

What's tricky is when you've got plural demons and they both possess something: *the two demons' victim*. (Unpacked: *the [one] victim of the two demons*.) The apostrophe comes after the plural to show joint ownership. In speech, *demons* and *demons'* sound the same, but in writing, the apostrophe goes after the *-s* to show plurality and possession.

Demons, *demon's*, *demons'*: these are the three inflections of the base noun to signal the plural, the possessive, and the plural possessive.

Verbs, adjectives, and adverbs similarly inflect: we'll cover these forms in Lesson II.

Derivations do change the lexeme of the word, thereby creating a separate entry in the dictionary. Add certain morphemes on either end of the word—or both ends—and you change the word from a noun to an adjective, or negate the word, or turn it into a bloated nominalization (this is the danger of zombie nouns). Some examples:

demonize, verb. To portray as wicked and threatening.
demonlike, adverb or adjective. Like a demon.

demonic, adjective. Of, resembling, or characteristic of
   demons or evil spirits; fiercely energetic or frenzied.
demonically, adverb. In the manner of a demon.
demonolatry, noun. The worship of demons.
demonology, noun. The study of demons.
demonologist, noun. A person who studies demons or
   demonic belief.
dedemonification, a noun I made up. The act of ridding
   oneself of a demon (see also *exorcize*).

Compounds form a kind of derivation that combines more than
one root morpheme. Sometimes they're jammed together with no space
("closed up"). Sometimes they're hyphenated. Sometimes they're separated with a space. I make sure, when I'm writing and editing, to look
up which one it is using *Merriam-Webster Online* or the *OED*. Examples of compounds:

demonkind, noun. Demons collectively.
demonland, noun. A realm inhabited by evil beings or
   demons.
demon star, noun. Algol, the second-brightest star in the
   constellation Perseus.
demon bowler, noun. A really great bowler.
demon cook, noun. A really great cook.
demon-born, noun or adjective. Born of demons.
noonday demon, noun. The sin of sloth.
demon-weary, an adjective I made up. Weary of the topic of
   demons.

I shut Volume D. This dive into one headword confirmed what I
already knew: Lexemes are legion. The edges are fuzzy. At the end of
the day, it's a miracle any of us communicate on any level at all.

Definitions change over time, as I saw with *demon* and, in general, I believe we should let them. We can also choose to drop old word usages that exclude or deride specific groups of people. As an editor, I am very tired of changing *he* to *they* when editing STEM textbooks—but I'm even more tired of being subtly excluded on account of my gender. So I will keep changing *he* to *they*, *manned* to *crewed*, *man-made* to *artificial*, *synthetic*, *fabricated*, or *manufactured* (from *manu*, Latin for hand), and *mankind* to *humankind*.

Indeed, whenever I see the word *mankind*, I think of Jack Handey's "Deep Thought" from a 1993 broadcast of *Saturday Night Live*: "Maybe in order to understand mankind, we have to look at the word itself. 'Mankind.' Basically, it's made up of two separate words—*mank* and *ind*. What do these words mean? It's a mystery, and that's why so is mankind."

Using *they* instead of *he* (or the awkward *he or she*) is now used by *The Washington Post* and many other publishers. As much as it pains the more prescriptive among us, if Shakespeare and Jane Austen used *they* to refer to one person, so can we. Subatomic changes can make enormous differences to readers.

UNCHOOSING WORDS: BEING CONCISE

Word choice is a skill you can develop by applying the information in this lesson, but word choice also means unchoosing words the reader doesn't need. In ancient Greece, people from Laconia—the Spartans—were famous for being laconic, speaking in concise, blunt statements. They were also famous for cutting people up. In writing, we're told to be concise (from Latin, *concīsus,* to cut up), but how exactly do you do that, without lacerating or mangling your hard-won words?

Joseph M. Williams, in one of my favorite style guides, *Style: Lessons in Clarity and Grace* sums up concision in six steps:

1. Delete words that mean little or nothing.
2. Delete words that repeat the meaning of other words.

3. Delete words implied by other words.

4. Replace a phrase with a word.

5. Change negatives to affirmatives.

6. Delete useless adjectives and adverbs.

When you go through your draft, make sure every word deserves its place. Extra noise in your message is the static that interferes with your message to the reader. The Golden Rule applies not just to human behavior but also to writing: write unto others as you would have them write unto you. If you take time to interact with the English field of words and form words your intended audience will most likely appreciate, then you'll be sending out good vibes to the reader, one -*eme* at a time.

EXERCISE 1

Build your mental lexicon by creating a Word Well, in a notebook or a spreadsheet. Read the dictionary for at least ten minutes—any dictionary, from the Scrabble Dictionary to *Merriam-Webster Online* (http://merriam-webster.com/browse /dictionary/) to the *Oxford English Dictionary*. If you're a student, you can usually access the *OED Online* through your university library for free. Start at the beginning, or in a random place, or with your favorite letter. Write down at least ten particularly vibrant words that evoke something sensory for you, a word your primary audience would like. What kind of words are you drawn to, what kinds of sounds? If you have a writing topic in mind, find words that will help readers understand your topic. Optional: Post a few favorite (or curious!) words and their definitions to Twitter with the tags #SubatomicWriting and @jamiezvirzdin.

EXERCISE 2

Look up the etymology of your ten words. Google's dictionary—and Wiktionary, Wikipedia's free dictionary—have quick, decent etymologies; the *Oxford English Dictionary* has a delight-fully comprehensive one. Are your words mostly Germanic? Latinate? French? Greek? A mix? From somewhere else entirely? If the words you chose are nominalizations or are not suitable for your primary audience, can you reduce the number of morphemes or find a more common synonym? Optional: Post a few favorite etymologies to Twitter.

EXERCISE 3

Take one sentence from your science writing. Label the words according to the abbreviations listed in Table 0.1: *lex* for Lexical words, *func* for Function words, etc. Or, if you're feeling whimsical, use the quark abbreviations, like *u* for Up, *d* for Down. What's the balance in your sentence? Then, take a favorite author's sentence. What balance of Lexical and Function words, common or complex, do they use? Go back to your sentence and cut out unnecessary words according to the six principles of concision from Joseph M. Williams. Optional: Do the same for a whole paragraph.

NESTED CLASSES

Phrase Level

I like crossing the imaginary boundaries people set up
between different fields—it's very refreshing. . . . It's about
being optimistic and trying to connect things.

—Maryam Mirzakhani, mathematician, first woman and first Iranian
to win the Fields Medal, 1977–2017

IN THE PREVIOUS LESSON, the Higgs boson and the
Higgs mechanism represented word choice and the process by which
words form and take on meaning. In this lesson, we take a few fuzzy,
vibrating words, assign a word class to each, and glue them into a
phrase, a small group of words that forms its own conceptual unit in
English grammar.

A stable phrase has the power to convey much more than the sum
of its individual words. It's not a complete sentence yet—it's not even
the dreaded clause. Each phrase is a fragment, like when you give a
quick response or reveal the murderer in the boardgame Clue: *The demon. Blue. In the library. Took my cat. Hurrying to write the full book
before she comes back. Very mean of her.* Writers sometimes rush the
phrase phase of the writing craft, but I recommend slowing down: hasty
gluing makes for a poor final product.

To help us at the crafting table, we turn to our second boson in the
physical universe, the **gluon**. Gluons are massless particles that bind

quarks together, usually in twos and threes, as Dyna mentioned, but sometimes more. The result is a *composite particle* whose mass and energy far surpass the sum of its quarks (we don't know why yet). The Higgs mechanism gives quarks and leptons a little mass, but the process of gluing quarks together is what gives us most of what we call common matter. In the literary universe, gluons represent **phrase rules**—sticky, complicated, fickle rules that govern what types of words can group together and in what order.

The most stable kinds of quarks, Up and Down, are glued into stable groups of three, creating our proton (Up + Up + Down, or **uud**) and our neutron (Down + Down + Up, or **ddu**). Each of these composite particles has a mass of one "atomic mass unit," or one **dalton**, named after our friend John Dalton, whose bowling habits inspired our ball-and-stick chemistry models. Without the glue-like stability of the individual proton or neutron in the nucleus of atoms, we wouldn't exist. There are eight kinds of gluons that use the **strong force** to bind quarks, and I've matched them to the eight major types of phrases in English that bind words. (We won't get to these until the end of the chapter, but for those who want a sneak preview, you can turn to Table 2.2.) The mapping is not pretty and not perfect, but neither is our understanding of the subatomic realm.

Gluons can also glue together less stable Charm, Strange, and Bottom quarks (which we left out of Table 1.1), as well as antiquarks like AntiUp, AntiDown, and the rest. Like a volatile Facebook group, however, these less stable composite particles don't last long. I know Dyna said not to mention antiquarks, but I must at least briefly mention some of their composite particles. **Mesons** (from the Greek *meso-*, intermediate) are unstable composite particles with two quarks: quark + antiquark. My favorite of these is the *pion*. **Baryons** (from the Greek *barýs*, heavy) have three quarks or three antiquarks. While protons and neutrons are the most stable baryons, the Lambda baryons, for example, have less stable combinations and therefore shorter lifetimes (**udc**, **uds**, and **udb**). After these triquarks come the tetraquarks, pentaquarks, hexaquarks, and other "XYZ particles." The Greek and Roman alphabet soup in particle physics is definitely an acquired taste.

As fun as these exotic particles are, they are not stable and not common, so when it comes to Subatomic Writing—writing that matters to the reader—we make sure to glue together stable, commonly used constructions, but a few exotic ones along the way are fine.

Glu was an Old French word for bird-lime—a sticky substance spread on twigs to trap birds. Rather than imagining three quarks trapped together in a sad, sticky mess, I picture the proton or neutron as one strong nest holding three eggs.

NESTS IN NESTS IN NESTS

When my Telescope Array colleagues and I drive into the desert to check on our scintillation detectors, we sometimes find raven nests atop the electronics boxes. (If there are eggs or babies inside, we wait until they've grown and flown before we remove the abandoned nest.) Raven nests are enormous and hearty structures, built with thick sticks and glued together with "white dielectric material," a gift the birds also leave on our solar panels. Some of these impressive nests are almost a meter across and could easily hold many smaller nests built by smaller birds.

In a sentence, words nest together in phrases, but phrases can also nest inside other phrases. The idea of nests within nests is called **recursion** in English grammar—and math, and computer programming, and quantum physics. A simple example of recursion in English is the sentence, "They said [that he said [that she said [that they said [that he said [that she said . . .]]]]]." You could go on forever and never hear what the fuss was about. If you picture a sentence like a long hallway filled with doors, and every door leads to a hallway lined with more doors, you must remember to return to the original hallway and lead your reader to the exit before they die of starvation, boredom, or confusion.

Sometimes the nested phrases in a sentence—particularly in science writing—become too much like the worlds-within-worlds setup in the movie *Inception*, so good writers learn how to unpack the nests, setting them next to each other instead of grammatically folding them inside one another. Other times, there are too many similar phrases in a row, and it's better to nest them, simplify them, or divide the thought into two separate sentences. And if the nest isn't needed anymore, abandon it.

Gluons can also interact with themselves, as phrases do: a gluon can form another gluon; two gluons may combine to form one, and gluons may even reduce—unnest, as it were—to pairs of quarks (these "glueballs" have yet to be confirmed). In the end, recursion is a remarkable feature to express complex thoughts, but it can quickly turn into a nested nightmare for readers.

WARNING: BORING STUFF AHEAD

Don't try to read this chapter for, like, fun. I suggest you skip to the section of the word class that you have the most trouble identifying. If you don't understand that section, maybe back up in the text until you do. Even though this lesson might feel like a slog, I've been thorough because, believe it or not, most problems in punctuation and in grammar can be traced back to an incomplete understanding of word class and how phrases are put together. Be patient, take it slowly. Try to find beauty in the pain?

WORD CLASS

The first step in becoming a nest-master is conquering **word class**. Also called *parts of speech*, this verbal taxonomy describes what role the word is playing in a phrase. What's tricky is that the same word can switch word classes. A noun in one phrase is a verb in another phrase. For example, take the word *well*:

Noun (Actor): The *well* of words runs deep.
Verb (Action): Tears *well* up in students' eyes during
    Lesson II.

Adjective (Modifier): All will be *well* by the end.
Adverb (Modifier): Are things going *well?*
Interjection (Insert word): *Well*, fine then.

And it gets ugly fast, culminating in the many diverse, abstract uses of the word I dislike most in the English language: *that.* See the appendix for more about this terrible word.

The number of word classes is still hotly debated—is it three? Is it fifteen? Like one of those old 1980s dance aerobics shows, I'll give you Easy, Medium, and Advanced versions of word class. You choose the one that is best for you. Grab your leotard, leg warmers, and mullet: it's time for a brain workout.

Easy (Four Word Classes)

I've found that almost all Lexical and Function words easily sort into the roles of *actors*, *actions*, *modifiers*, or *connectors*. (We drop the whole family of Insert words.) To give you a definition and an example of each, I've italicized the example sentences, underlined the multi-word phrases that act as one unit, and used boldface for the word classes being described.

1. An **actor** does the action or receives the action: ***Dyna*** *took my* ***cat***. An actor can be a person, animal, place, thing, or idea. A few unusual actors have a special feature that allows them to be actions, too, creating room for more nests with more actors. To find the actors in a sentence, ask: Who is acting? Who is receiving the action, either directly or indirectly? Sometimes a less interesting understudy takes the place of the primary actors so the show will go on: ***She*** *took* ***him***.

2. An **action** is any activity the actor does: *Dyna* ***took*** *my cat.* Some actions need more than one word: *She* ***will return*** *soon.* Or the action equates an actor to something else. *She* ***is***

*a demon. I **feel** tired.* I like to mentally replace these kinds of verbs with equal signs. *She = a demon. I = tired.* To find the actions in a sentence, ask: What is the actor doing? Or, where is the equal sign?

3. A **modifier** changes an actor, an action, or another modifier in some way, describing or limiting it: *Dyna took **my** cat. She will return **soon**.* (Whose cat is it, specifically? When will she return, exactly?) If you can delete a word without losing the gist of the sentence, the word is probably a modifier, a prop in the scene. Some modifiers are phrases: ***in the library***. (Where did this disaster happen?) To find modifiers that describe actors, ask: What kind? Whose? What color? Which one? How many? To find modifiers that describe actions, ask: When? Where? Why? How? In what manner?

4. A **connector** links actors to actors, actions to actions, modifiers to modifiers, phrases to phrases, clauses to clauses: *this **and** that, sing **or** shout, **neither** yellow **nor** gold.* A second kind of connector makes one group of words less important than other words: ***If you can't say something nice***, *don't say anything at all. Don't move **unless** I do.* You know words are less important in a sentence if you can delete them and still have a complete thought that makes sense: *Don't say anything at all. Don't move.*

In our earlier headline "Blue Demon Nabs Cat Yesterday," we'd easy-tag the word classes as modifier, actor, action, actor, modifier. If we abandoned the modifiers, we'd still be left with a complete headline, "Demon Nabs Cat" (actor, action, actor).

In Subatomic Writing diagrams (see the appendix), you will generally find actors and actions on horizontal lines; modifiers on diagonal lines under the words they modify; and connectors on vertical dotted lines, with the second kind of connector on slanted dotted lines.

Medium (Eight or Nine Word Classes)

In 1867, a physician and traveler named A. W. Chase published a book of recipes and instruction sets. He titled it *Dr. Chase's Recipes; or, Information for Everybody: An Invaluable Collection of About Eight Hundred Practical Recipes, for Merchants, Grocers, Saloon-Keepers, Physicians, Druggists, Tanners, Shoe Makers, Harness Makers, Painters, Jewelers, Blacksmiths, Tinners, Gunsmiths, Farriers, Barbers, Bakers, Dyers, Renovators, Farmers, and Families Generally.*

Curiously, among the recipes for how to cure ingrown toenails, create a varnish for a steel plow, prevent smallpox from "Pitting the Face," distinguish the parts of a horse, and fry eggs "extra nice," there was a little poem about grammar, a great introduction to the jargon of word class (Figure 2.1).

"For with these lines at the tongue's end," Chase says, "none need ever mistake a part of speech." While many of his other recipes are now obsolete (and grounds for malpractice), the poem generally holds true. There are eight parts of speech according to *The Chicago Manual of Style*, 17th Edition (*Chicago*, Section 5.3):

1. nouns
2. pronouns
3. adjectives
4. verbs
5. adverbs
6. prepositions
7. conjunctions
8. interjections

(*Chicago* groups articles, which the *Longman Student Grammar of Spoken and Written English* calls *determiners*, with adjectives.) Kory Stamper, who worked as a lexicographer for *Merriam-Webster*, takes

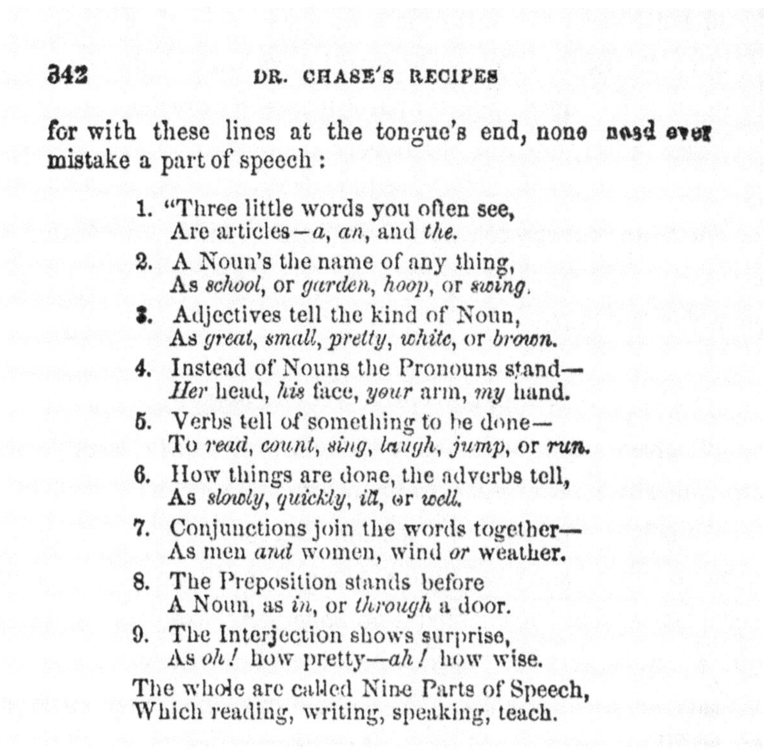

342   DR. CHASE'S RECIPES

for with these lines at the tongue's end, none need ever mistake a part of speech :

1. "Three little words you often see,
   Are articles—*a, an*, and *the.*
2. A Noun's the name of any thing,
   As *school,* or *garden, hoop,* or *swing.*
3. Adjectives tell the kind of Noun,
   As *great, small, pretty, white,* or *brown.*
4. Instead of Nouns the Pronouns stand—
   Her head, *his* face, *your* arm, *my* hand.
5. Verbs tell of something to be done—
   To *read, count, sing, laugh, jump,* or *run.*
6. How things are done, the adverbs tell,
   As *slowly, quickly, ill,* or *well.*
7. Conjunctions join the words together—
   As men *and* women, wind *or* weather.
8. The Preposition stands before
   A Noun, as *in,* or *through* a door.
9. The Interjection shows surprise,
   As *oh!* how pretty—*ah!* how wise.

The whole are called Nine Parts of Speech,
Which reading, writing, speaking, teach.

FIGURE 2.1 A "Grammar in Rhyme."

Source: A. W. Chase, *Dr. Chase's Recipes* [. . .]. Ann Arbor: A. W. Chase, 1867, p. 342.

word classes back further in time—to ancient Greece, of all places—in *Word by Word: The Secret Life of Dictionaries:*

The parts of speech we use today were established in the second century [BCE] in a treatise called *The Art of Grammar* [attributed to Dionysius Thrax, a Hellenistic grammarian], which gives us our first incarnation of the eight parts of speech: noun, verb, participle, article, pronoun, preposition, adverb, and conjunction. This system has been futzed with over the centuries: article was dropped, interjection was added, participle was later considered a flavor of verb, and adjective was pried out of the noun class and became its own thing. By the time English lexicographers came on the scene in the late Middle Ages, our parts of speech were fixed and based entirely on Latin and Greek.

The truth is that word classes are hard to pin down because the boundaries are fuzzy—in part because, as we saw in Lesson I, words themselves are fuzzy.

Advanced (Twelve Word Classes)

For the adventurous, we examine the following word classes: nouns, verbs, adjectives, adverbs, determiners, pronouns, primary auxiliaries, modal auxiliaries, prepositions, coordinating conjunctions, subordinating conjunctions, and finally, our old friends the Insert words. If these twelve classes that I've adapted from *Longman* make your heart sink, don't worry. Writers don't need to be linguists to write well. However, the faster you can distinguish a raven egg from an interloping cowbird's egg, the easier writing, revising, and editing become.

Looking back at Table 1.1, we see four word classes listed under common Lexical words, seven word classes listed under common Function words, and of course, the chatty Insert words, including interjections. This is how we will divide and conquer each word class, and then we will glue these word classes together into eight types of phrases.

Up Words (Lexical Words)

Don't forget to take a break—or a nap—if your brain needs time to digest new concepts. I recommend copying the grammar terms into a physical notebook and adding your own example of it. Mixing up handwritten flashcards and quizzing yourself also helps. (Research shows handwritten notes and drawings help students learn and remember material better than typing out notes.) I often invite my students to color-code these twelve word classes with whatever colors they like.

Speaking out loud or signing the word as you go also increases the like-lihood you'll remember the information.

1. NOUNS (n)

Semantically—at the level of lexemes, as introduced in Lesson I—**nouns** refer to physical or abstract entities (or both, if you're talking about invisible subatomic particles). Nouns can be **common** (*cat, book, physics, information*) or **proper**, where you capitalize the first letter of a specific person, place, or thing (*Jamie, Greece, October*).

Morphologically—at the level of morphemes—you may remember that nouns can change by adding **inflectional suffixes** to the end of the noun. To change a noun's **number** from singular to plural, we usu-ally add an *-s* or *-es* (*two demons, five million octopuses*). Some **noncount** nouns can't have a plural form (*information, pollution, water, happi-ness*). And then there are pesky **irregular plurals** (*mouse/mice, child/children, datum/data*, although *datum* has fallen out of general us-age) that you just have to memorize. Google "Irregular plural nouns in English," and you'll find plenty.

The **possessive** noun, remember, makes the noun own something: *a demon's victim*. You can combine the plural and the possessive into the Dyna-awful **plural possessive**: *the two demons' victim*. These "pos-sessive nouns" act like adjectives, modifying another noun or noun phrase. In fact, they are diagrammed as adjectives in the Reed-Kellogg system of sentence diagramming.

Chicago (Sections 7.16–7.22) says to keep this possessive and plu-ral system the same for all nouns, abbreviations of nouns, and numbers used as nouns, even if the words end in *s*, *x*, or *z*: one ***bass's*** *stripes, two* ***puppies'*** *paws,* ***Dickens's*** *novels,* ***Jesus's*** *adherents,* ***Borges's*** *library, the* ***Lincolns'*** *marriage*.

We've already discussed **compound** nouns (*noonday demon*) and nouns with **derivational suffixes** (*demonic, friendliness*) that turn the noun into a different word class. These changes can cause the word to mean something quite different (compare: *friend*, **girl-friend**, **ex**-*girlfriend*).

2. VERBS (v)

Lexical verbs are the central actions in our sentences. They describe an action (*to stop*, *drop*, *roll*), a process (*to build*, *study*, *destroy*), or the state of something (*to exist*, *feel*). Science writers shoot for lively, vivid, short, old Germanic verbs whenever possible.

The **primary verbs** *to be*, *do*, and *have* are the most common verbs in English. As main verbs, primary verbs are powerful, but they can also muddy meaning; use them wisely. *Be*, *do*, and *have* also act as primary auxiliary verbs (helping verbs), which are so essential and so helpful that they get their own "p.aux" word class under Function words.

The most fundamental form of any verb is its **infinitive**, *to* + the **base**: *to be*, *to live*, *to laugh*, *to love*. *To* is not a preposition here: it marks the infinitive and is literally called "the *to*-particle." Is it OK to "split the infinitive," *to **boldly** put* a word in between the *to*-particle and the base? Yes. This is a dumb, Latin-based rule (Latin infinitives are one word) that has no bearing on a Germanic-based language. The English infinitive is much more flexible.

By itself, the infinitive is like a car parked in neutral: *to drive*. To put the verb in gear, drop the *to*-particle and keep the base: *drive*. You've now **inflected**, or **conjugated**, the verb, changing it in some small way to let the reader know who will be driving at what point in time. Now it's a **main verb** and revving to go.

A main verb has five usual ways it can conjugate—five gears, to use a car analogy—so buckle up. Verbs shift to accommodate **person**, **number**, **tense**, **aspect**, or **voice**:

1. **person**: *I **make**, we **make*** (first person). *You **make**, you (all) **make*** (second person). *She/he/it **makes**, they **make*** (third person). Stay with that person from paragraph to paragraph. If

you are writing a science essay or a sci-fi story in third person, for example, don't suddenly switch to first person unless you're going for a braided form with clear section breaks.

2. **number**: *I am* (singular). *We are* (plural). Watch out for using a plural noun but a singular verb, or vice versa. This happens especially when the noun is far away from the verb in the sentence.

3. **tense**: present or past, now or then. *I write*, *I wrote*. *I lie on the couch* (present tense). *I lay on the couch* (past tense). *I lay the book down* (present tense). *I laid the book down* (past tense). Writers often switch tenses on accident, so check your paragraphs for shiftiness. Some other common writing issues related to tenses:

   - **"third-person singular simple present tense"** (base + -*s*): *She **writes** about physics*. This is one of the few present-tense verb conjugations that the English language clings to. But, if you're using *they* as a singular, gender-neutral pronoun, use: *They **write** about physics*.
   - **"future simple tense"**: *I **will write***. We signal a future event by keeping the base verb and adding *will*, a modal auxiliary verb (which also has its own "m.aux" word class in Function words).
   - **"future progressive tense"**: *I **will be writing***. We signal something we're about to do with the modal auxiliary *will* plus the primary auxiliary base *be*, and we put those in front of the "present participle" form of the verb, which adds -*ing* to the base: *writing*. If the verb ends in an *e*, it must be cut off before adding -*ing*.
   - **"future perfect progressive tense"**: *I **will have been writing***. We signal something we plan on completing in the future by combining one modal auxiliary and two primary auxiliaries: *will* + *have* + *been*. At the end, add the present

participle of the base, which ends in -*ing*. Obviously, avoid this mouthful when you can, but it's there if you need it.

4. **aspect**: the way an event spreads over time. Aspect describes if an action is instantaneous or timeless (*he eats*), continuous or **progressive** (*he is eating*), completed or **perfect** (*he ate*), or both—something ongoing in the past, the **perfect progressive** (*he was eating*). If you're speaking in the past tense and you want to refer to a completed action that happened even *further* back in time, use the **past perfect**: *He **had eaten** before he **noticed** the cat was gone*. Finally, aspect can also describe something **habitual**, something that carries on from the past into the present (*he has eaten*). Writers accidentally switch between these all the time.

5. **voice**: voice refers to the two verb-driven ways to arrange the actors, actions, and the acted upon in a sentence: who is doing what (active voice) or what is being done by whom (passive voice). The sentence *Dyna **made** a mess* is written in active voice: Dyna is doing something to something else. Passive voice is when you unnest Dyna as the subject of the verb (the actor doing the action) and tack her to the end of a prepositional phrase with the preposition *by* (that is, if you want to keep her at all): *A mess **was made** by Dyna*. The verb changes from the active voice *made* to the passive form *was made*.

For the record, passive voice is fine in scientific writing when you want the reader to keep track of an *object* rather than *who* is doing something to that object. But you can also use vivid verbs in active voice to make your abstract object more relatable. Compare:

- *Subatomic particles **smash** together in particle accelerators* (optimal: the abstract noun acts in active voice; the focus is on the particles).

- *Physicists **smash** subatomic particles together in particle accelerators* (also active, but with more focus on the physicists).
- *Subatomic particles **are smashed** together in particle accelerators <u>by physicists</u>* (passive, focused on the particles, but less engaging for the reader to read).

Therefore, passive voice is fine when you need it, but do avoid putting an active verb and a passive verb in the same sentence.

Finally, there are three other kinds of variation seen in the structure of verb phrases:

- **modality**, whether or not the verb is used with a modal verb like *will*, *can*, or *might*. These modal auxiliary verbs signal possibility or necessity and come before the base.
- **negation**: *she writes* versus *she does **not** write* (*not* usually acts as an adverb)
- **mood**: indicative, imperative, subjunctive (More about mood in Lesson III.)

In Table 2.1, you'll notice that the verb phrases near the top of the chart are what you'd see in a kid's book. The phrases at the bottom of the chart, in contrast, sound very clunky. These are places where you can revise for clarity and simplicity.

What can help you control where your verb goes is knowing the difference between regular verbs and irregular verbs. **Regular verbs** are verbs whose past tense and past participle are formed with *-d* or *-ed*: *I **jump** / I **jumped** / I have **jumped**. I **try** / I **tried** / I have **tried**.*

Irregular verbs are tougher: *She **bites** me / She **bit** me / She has **bitten** me. You **swim** / You **swam** / You have **swum**. He **lies** on the couch / He **lays** on the couch / He has **lain** on the couch. They **lay** the books down / They **laid** the books down / They have **laid** the books down. We **let** it happen right now / We **let** it happen yesterday / We have **let** it*

TABLE 2.1
A handy-dandy verb chart.

ASPECT AND/OR VOICE	PRESENT TENSE	PAST TENSE	MODALITY
simple	writes, write	wrote	could write
progressive	am/is/are writing	was/were writing	could be writing
perfect	has/have written	had written	could have written
passive	am/is/are written	was/were written	could be written
perfect + **progressive**	has/have been writing	had been writing	could have been writing
perfect + passive	has/have been written	had been written	could have been written
progressive + **passive**	am/is/are being written	was/were being written	could be being written

Note: The more verbs and verb forms you know, the stronger your writing game can be.

happen again and again. I'm afraid you just have to google "List of ir-regular English verbs" and review or memorize them.

A final note on verbs: some **phrasal verbs** have multi-word verb forms (*break down, come down with, drop in/by/over*). These are extra particles, literally called "adverbial particles," that come after the verb, like a truck hauling a camper or horse trailer. Others are **derived verbs**—verbs made from nouns or adjectives using morphemes (*soften, clarify, radicalize*). Be careful, because these aftermarket parts on your verb car can weaken your verb.

English verbs take time to understand, but you'll be an expert mechanic before you know it.

2.5. VERBALS: The PIG

Surprise! I'm deviating from *Longman* to talk about verbals, monsters, and Minecraft.

I think Minecraft is a great computer game. I'm glad Max likes it, but I seriously dislike piglins, who are half-human, half-pig characters controlled by the computer in the area of the game known as the Nether. Piglins can help or hurt you—barter with you for items or murder you. They freak me out. In like fashion, before we leave nouns and verbs, we must face three similar monsters in grammar: the **participle**, **infinitive**, and **gerund** (**PIG**, for short). These are the **verbals**, and they cannot fit neatly into just one word class.

I mention the PIG now because each of these grammar monsters are half verb and half something else, and two of these three monsters can become half verb, half noun. We'll cover the three monsters simply at first, as one-word concepts. Later, in the section called Verbal Phrases: The PIG Strikes Back, we meet the *Participle Phrase, the Infinitive Phrase*, and the *Gerund Phrase*. I prefer even Dyna to these nightmares.

The **participle** is half a verb, a parasite, a vampire: it can't exist on its own. It doesn't even work as a good fragment most of the time. It is born from the infinitive form of a verb (like *to barter* or *to murder*). Ill-begotten, the participle forms either its present or past incarnation, usually with *-ing* or *-ed*: **bartering, murdered**. Then, it feeds, as you just saw in the verb section: the participle becomes the main verb as it hooks up with primary auxiliary verbs and modal auxiliary verbs: *The piglin is **bartering** with me. I could have been **murdered** by the piglins.* We will see more of present and past participles later.

But wait, it gets better: a participle can also abandon its auxiliaries and instead attach itself to a noun, becoming half verb, half adjective: the **participial adjective** (present or past). This monster can come before or after the noun it modifies. *Look at that piglin **bartering**. The **murdered** piglin dropped its axe.* You can identify the participle in this form by asking the same questions you'd ask of modifiers: What kind of piglin is it? The bartering one. The murdered one.

We met the **infinitive** earlier, as a neutral starting point for conjugating verbs, but here's the truth: the infinitive is the primordial mother of all verbs, like the Greek monster-goddess Gaia, whose Wikipedia page reads like super disturbing erotica. Remember, this monster-goddess is usually preceded by her herald, the **infinitive particle** *to*: *to mourn, to laugh, to burn.*

Sometimes this base verb hooks up with a main verb, and she becomes a **complementary infinitive**: *The piglins <u>dared **to attack**</u>. <u>Are you **able to fight**</u>? You <u>**ought to run**</u>.* Other main verbs that willingly take a complementary infinitive include *begin, learn, continue, remember, forget, be accustomed, cease, hesitate, undertake,* and *fear.*

At other times, this monster-goddess becomes half verb, half noun: *I want [to win].* Here is a non-piglin example by Annie Leclerc: "***To laugh*** *is **to live** profoundly*" (as quoted in *The Book of Laughter and Forgetting* by Milan Kundera and translated by Aaron Asher). This is the **nominal infinitive**, an infinitive used as a noun. (If you can name something, you can conquer it.)

The infinitive also transforms into half verb, half adjective, attaching herself to a noun: *Now is the right <u>time **to scream**</u>.* This is the **adjectival infinitive**.

Or she becomes half verb, half adverb, modifying the verb or Verb Phrase in the sentence. You can identify this **adverbial infinitive** by placing the words *in order* before the infinitive: *(In order) **To win**, you <u>must barter and murder</u>.* You can also ask why: Why must you barter and murder? To win.

Worst of all, sometimes the infinitive ditches her particle: *I dare not [to] **follow**. Help me [to] **survive**. Let me [to] **go**. Make them [to] **stop**.* She becomes the **bare infinitive**.

The **gerund** is half verb, half noun. I call him Gerald. He's the least of our troubles. How do you create him? Add *-ing* to a verb and use it as a noun: ***Bartering*** *is great. I like **murdering**. Longman* prefers talking about "*ing*-form" and "*ing*-participle" and "*ing*-clause," refusing to mention either gerund or Gerald, but I like to keep it simple. If you can replace the *-ing* word with another noun and the sentence still makes sense, you've found Gerald. If it doesn't make sense, you've got a vampire to deal with.

Stay tuned for more as the PIGs start to nest.

3. ADJECTIVES (adJ)

An adjective modifies a noun. The adjective *blue*, for example, limits or further describes the noun *demon*. Which demon? The *blue* one.

You can change the degree or the intensity of adjectives that are **gradable** (*darker, highest, more fun* and **most fun**, but *funnier* and *funniest*). Adjectives generally take the inflectional suffixes *-er* for the **comparative** form (*cheaper*) and *-est* for the **superlative** form (*cheapest*). You can change a noun or verb into a **derived adjective** with various suffixes (*reachable, eventful, professional*). It is best to recognize when you are using excessively derived adjectives; change them back to nouns and verbs (dezombify them). For a full list of derived adjectives, google "Adjective suffixes" to see all the morphemes that turn words into adjectives.

Sometimes adjectives act like nouns (they want to be cool, too). These we call **substantive adjectives** or **adnouns**. Example: *The **old** [people] and the **young** [people]. The **best** [students] and the **brightest** [students]*. The nouns are left out because the speaker presumes that the audience understands the nouns from context.

Attributive adjectives come before the noun they modify: <u>The **lucky** penny</u>. Other adjectives come after a verb: <u>The penny is **lucky**</u>. These adjectives are **predicative adjectives**—that is, they are found after the verb.

Compound attributive adjectives cause many <u>**hyphen-allergic** writers</u> to run into trouble. Removing the adjective nearest the noun will help you know if you need a hyphen: *hyphen writers* doesn't make sense. Remove the other adjective: *allergic writers* doesn't make sense either, in the context of the sentence. But *hyphen-allergic writers* does make sense: we understand that it means *writers who despise hyphens* (and that's OK).

Participial adjectives—as an extra review to spot those vampiric verbals—are adjectives formed from verbs. There are two types. First comes the **present participial adjective**: *The flying superhero acci-dentally ran into the billboard.* What kind of superhero? The flying kind. The **past participial adjective** describes action in the past and often takes an irregular form: *Hers was a city torn by strife, a city of fallen angels and broken promises.*

The correct order of adjectives comes naturally to native English speakers, but even native speakers can stumble if there's too many adjectives in a row (something we generally want to avoid). The order of adjectives is:

1. **Determiner** (*a, the, this, our, every, what, three, no*)
2. **Observation or Opinion** (*fantastic, gorgeous, weathered*)
3. Physical description: **Size** (*huge, tiny*)
4. Physical description: **Shape** (*circular, thin*)
5. Physical description: **Age** (*ancient, juvenile*)
6. Physical description: **Color** (*blue, yellow*)
7. **Origin** (*Nicaraguan, American, Marshallese*)
8. **Material** (*silk, gold, metal*)
9. **Qualifier or Purpose** (*coat*, as in *coat rack*; *running*, as in *running shoes*)
   (Note: Sometimes Shape and Age are switched.)

You can identify adjectives by answering the following questions (taken from the order of adjectives). Let's assume we've identified the noun *bird* and are trying to find its modifiers.

Which bird is it? *the* bird
How many birds are there? *one* bird

What kind of bird is it (opinion)? *stately* bird
What size of bird? *enormous* bird
What shape of bird? *round* bird
What age of bird? *ancient* bird
What color bird? *black* bird
What is the origin of the bird? *Southwestern* bird
Of what material is the bird? *feathered* bird
What purpose does the bird serve? *flying* bird

As Drake Baer reported in *The Cut*, in an article titled "The Unexpectedly Existential Roots of Adjective Order," these adjective-ordering restrictions happen all over the world. One theory about the order regards *innateness*, that is, "the closer you get to the noun, the more the adjective speaks to the essential nature of that noun." The temperature of the Sun, for example, is more critical than your opinion of its pleasantness on a nice day. "It pleases the ear," Baer says, "to have adjectives placed like concentric circles, radiating out from the center of the noun."

4. ADVERBS (adv)

Adverbs modify an action, process, or state (that is, adverbs modify a verb, a half-verb PIG, or another adverb). They can describe the time, place, or manner of the action: *It happened **too quickly**. Dyna will be* ***here later this week***.

They can convey your attitude about your topic: *I will **absolutely** not stand for this*.

They can modify an adjective, even if the adjective itself is modifying a noun: ***too many*** *Twinkies*. You can start to see how one word modifying another word modifying another word creates recursion.

Adverbs are often formed from adjectives by adding the suffix *-ly*. Others don't have a special ending: *too, just, there, finally, however, usually*. Like adjectives, a few adverbs inflect to form comparative and superlative variations—or, if they already end in *-ly*, they use additional adverbs in their comparative and superlative forms: *soon/sooner/soonest*; *quickly/more quickly/most quickly*.

The **negator** *not* is an adverb-like particle modifying the verb (whereas the negator *no* usually functions like an adjective or an Insert word).

Although a hyphen is needed between compound attributive adjectives (*hyphen-allergic writers*), we do not use a hyphen after adverbs ending in -*ly*: *The **undeniably** evil demon is **unquestionably** mean*.

Horribly, we do use a hyphen with adverbs not ending in -*ly*—but only if they come before the verb: *A **well-known** physicist is **well known***.

Adverbs can also express a connection to what was stated earlier in a sentence, paragraph, or a preceding paragraph (these are often called **conjunctive adverbs**): ***Consequently,** we lost the battle*. These words are useful in connecting sentences and paragraphs together, leaving a coherent trail for the reader to follow (Lesson VI).

Many writers encourage other writers to use fewer adjectives and adverbs. I agree with this, to a point: whenever possible, turn adjectives into strong nouns, adverbs into stronger verbs. If you're describing something, this is harder advice to follow, but even then, you can vary what kinds of modifiers you tack onto nouns and verbs.

Down Words (Function Words)

5. DETERMINERS (det)

Determiners perform exactly like lexical adjectives, modifying nouns, but . . . they're more boring than lexical adjectives. They're functional and needed, and most of them are deeply Germanic, but they don't "pop" like lexical adjectives do. At the risk of making some linguists very angry, I generally call determiners adjectives, since determiners also modify nouns, noun phrases, and other adjectives. So whenever you see *determiner*, think "boring but necessary adjective." There are seven kinds of determiners:

1. **Articles**—*a*, *an*, *the*—tell the reader that a noun is about to appear. If it's any old noun, an unknown entity, or a member of a group, use *a* (or *an*, if the noun afterward starts with a vowel: ***an*** *apple*). This is the **indefinite article**. But if it's a known noun—known to the speaker and the primary audience—use the **definite article**, *the*: ***The*** *demon of this book*. Sometimes it doesn't really matter which article you use, and sometimes it does. And sometimes you can delete the article without losing anything, which is a great way to be more concise. (If English is not your native language, you might hate article particles, and that's OK.)

2. **Demonstrative determiners**—*this*, *that*, *these*, *those*—signal if the upcoming noun is near or far from the speaker, either physically or ideologically. They're like demonstrative pronouns, but they describe a noun instead of replacing it: ***This***/***that*** *class is amazing. I hate* ***these***/***those*** *darn demonstrative* *determiners*. Take out other modifiers to see what is modifying what: the sentence still makes sense if I say *I hate* ***these*** *determiners*, so you know that *these* modifies *determiners* directly. This trick is useful when you're diagramming sentences.

3. **Possessive determiners**—*my*, *your*, *his*, *her*, *its*, *our*, *their*—are modifiers used to indicate possession or ownership: ***Our*** *football* *team is better than* ***your*** *team*. ***Its*** *breath was fetid*. Another hang-up in English is mistaking the possessive adjective *its* (which, by all rights, should have an apostrophe like other possessive nouns) for the contraction *it is*. As my great Writing Fellow teacher in college, Beth Hedengren, liked to sing to the tune of "Row, Row, Row Your Boat": "I T apostrophe S always means *it is*! I T S shows possession like *my* or *her* or *his*."

4. **Indefinite determiners** (also called **quantifiers**) are similar to indefinite pronouns in that they refer to fuzzy, unspecified quantities, but they're used in front of nouns: *Most people prefer a few real friends over many superficial acquaintances.* You can use Google to list them all for you.

5. **Interrogative determiners** ask questions by requesting specific information, or they point to a specific noun: *In what universe does he live? Whose book did you buy? I didn't see which books they bought.* Interrogative determiners belong to a special category of words called **wh-words**: *who, whose, whom, what, which, when, where, why, how,* plus *whoever, whosever, whomever, whatever, whichever, whenever, wherever, whyever,* and *however.* But only three of them—*what, whose,* and *which,* plus their -*ever* counterparts—can act as interrogative determiners.

6. **Numerals**: *One fish, two fish.*

7. The **negative determiner**, used to negate a noun: *We'll be there in no time.*

Sometimes determiners can act like nouns (like substantive adjectives, they want to be cool, too). These we call **substantive determiners** or **adnouns**. Example: *Which seat is mine? The fourth [one]. How many apples does she have? She has four [apples].* The speaker presumes the audience understands what noun was dropped.

6. PRONOUNS (pron)

Pronouns are Function words that replace a noun or even a whole noun phrase. They are nouns, not modifiers. There are nine kinds of pronouns:

1. **Dummy pronouns** are—technically, according to *Longman*—oddities deserving of a word class all their own. Rather than have a grab-bag class, I've slotted them into our

existing categories, according to how they're most often used. Dummy pronouns include *there* and *it*:

- Also called **existential there**, this dummy pronoun announces something: *There are eight gluons*. Since *there* is not really functioning as an adverb (as in, "Ho! You there!"), it's often diagrammed by itself on a lonely line above the sentence *eight gluons are*. You could rewrite it: *Eight gluons exist*. We don't like that construction as much, however. I don't, anyway.

- The dummy *it* is also called **an extraposed subject**. *It* doesn't mean anything in the following sentence: *It is <u>necessary</u> **to write quickly**. There's no antecedent—no previous word—that the pronoun *it* refers back to. This is how you unnest an extraposed subject: ***To write quickly** is <u>necessary</u>*. Academics and science writers overuse the extraposed subject to sound more formal and important, but it can have the overall effect of distancing the reader. When possible, kill these two dummies and reach for stronger verbs. **Personal pronouns** act like nouns and refer to the speaker, the audience, or any other noun or noun-like entity. They are the most common kind of pronoun: ***I** can't force **you** to read **it***. Personal pronouns can be **subject pronouns** (*I, you, he/she/it, we, you all, they*) or **object pronouns** (*me, you, him/her/it, us, you all, them*), depending on where they sit in the sentence. A great way to tell the difference is to ask, "Who pushed whom into whom?" Subject pronouns go in the "who" position. Object pronouns go in the two "whom" positions. This can also help with the tricky compound constructions "She and I" (subject) and "her and me" (object), as in: *[**You** and **I**] pushed [**him** and **her**] into **them***. If you ever get confused, push out one of the pronouns from the nest and your ear will help you: *Me pushed he into they* doesn't sound right, but *I pushed him into them* sounds nicer, even if it's not very nice to push people.

2. **Demonstrative pronouns** are entities either closer to the speaker, further away (in time or idea) from the speaker, or farther away (in physical space) from the speaker: ***This** is amazing. I hate **those**.* They stand as nouns on their own; they don't modify a noun like demonstrative determiners do.

3. **Reflexive pronouns** refer back to a previous noun phrase, usually the subject of the clause: *I saw **myself** in the mirror. She made **herself** comfortable.* They can also intensify a statement: *I **myself** saw the demon.* However, just as you wouldn't say, *Myself discussed the topic,* don't say, *The students and myself discussed the topic.* Rather, use *The students and I discussed the topic.*

4. **Reciprocal pronouns** are like reflexive pronouns, but they show a mutual relationship with a previous noun phrase: *They came to respect **each other**.*

5. **Possessive pronouns**—*mine, yours, his, hers, ours, theirs*— are nouns (there is no possessive pronoun for *it*). They are closely related to possessive determiners like *my, your,* and *her.* Possessive pronouns include the meaning of the head noun, which may be absent from the sentence itself: ***Ours** is better than **yours**.* They stand on their own, unconnected from other nouns.

6. **Indefinite pronouns** are words that have a broad, fuzzy meaning. Some are quantifiers that stand by themselves (*all, many, some, most*), but others are compound words with a quantifier plus a general noun (*something, nobody, everyone*): ***Many** would show up to the party, but **nobody** would stay.*

7. **Relative pronouns** (*wh*-words, including *that*) are words
that introduce a relative clause (also called **relativizers**),
which we'll talk about in Lesson III: *The woman **who** lived in
Maryland had an experience **that** changed her life, **which**
profoundly affected her attitude toward demons.* These
pronouns stand on their own as nouns.

8. **Interrogative pronouns** (*wh*-words, but only *who, whose,
whom, what,* and *which,* plus their *-ever* counterparts) ask
questions: ***What** is going on here? Can you tell me **who** she is?
Whom did she push?* They stand on their own as nouns
(unlike interrogative determiners, which modify nouns like
adjectives do). Like interrogative determiners, interrogative
pronouns belong to the group of *wh*-words.

7. and 8. AUXILIARIES (p.aux and m.aux)

Auxiliary verbs, also called *helping verbs*, come in two flavors: **primary
auxiliaries** and **modal auxiliaries**. They are placed before a main verb
to "help" it—to show how the main verb is to be understood.

Primary auxiliaries: *Be, have,* and *do* are primary verbs, which
can function as both lexical verbs and auxiliaries.

Be, as an auxiliary, can form the **passive voice**: *The homework **was
eaten** by my dog.* (Active voice: *My dog ate the homework.*)

Be also forms the **progressive aspect**: *I **am** thinking of writing a
book.* Or, in the past tense: *I **was** thinking of writing a book.* (*Writing*
here is acting as a gerund.)

Have, as an auxiliary, forms the **perfect aspect**: *I **have** been here
before. I **had** not thought of that.* (*Not* is acting as an adverb here.)

Do is used to form negative statements and questions, called "the
do insertion": *It **doesn't** make sense. **Did** you see that? **Do** you understand that my cat's life is at stake here?* If you were to "unnest" those
questions, you would see a statement: *You **did** see that. You **do** understand that my cat's life is at stake here.* When diagramming a sentence,

you diagram the unnested question. *Do* is also used to intensify a noun: *I **do** want [to be a better writer].* (*To be* is an Infinitive Phrase acting as a noun here.)

Modal auxiliaries are used to express "modality," meaning possibility, necessity, prediction, and volition. My tenth-grade English teacher made us memorize the nine modals in a sing-songy voice: *can, could, shall, should, will, would, may, might, must.* (*Ought, dare, need,* and *had better* also sometimes act as modals). Fun fact: *would* was historically the past tense of *will.* Same with *can/could, shall/should,* and *may/might.* That time distinction is now less important than their personal and logical meanings:

- To express **permission, possibility, and ability**, use *can, could, may,* and *might.* Academic prose uses *could, may,* and *might* almost exclusively to express logical possibility. I was taught to use *can* or *may* if something was possible, but *could* or *might* to convey increased uncertainty, doubt, or tentativeness. (I checked, and this is still true.) Don't use too many of these "hedge" words when you're writing grants or writing about your research, but do be honest.

- To express **obligation**, use *should* and *must.*

- To express **volition or prediction**, use *will* and *would.*

Why should we even bother talking about this? These are very lovely, silk-fine nuances you can convey if you know this stuff. Passive aggressive? Maybe. Effective? At least sometimes.

9. PREPOSITIONS (prep)

Semantically, **prepositions** are Function words that stand at the head of **prepositional phrases**. They always signal the beginning of their own tiny nest.

Morphologically, prepositions don't change form. The best ones—the common ones—are short: *about, around, as, at, by, for, from, into, like, of, off, on, since, than, to, toward,* and *with.*

Avoid complex Function words, as mentioned in Lesson I. The following Strange prepositions can often be simplified: *such as, except for, apart from, regardless of, according to, by means of, on account of, in addition to, with regard to, as far as.* Why make a boring word more boring?

Are you allowed to end a sentence with a preposition? Yes, this was another dumb, Latin-based rule that doesn't need to be followed in English.

Like Mario and Luigi, Thelma and Louise, or Tweedledee and Tweedledum, prepositions are never without a buddy. The preposition's noun companion is called the **object of the preposition (OP)**. This companion—a noun, pronoun, gerund, or infinitive—stays close by in the sentence, usually appearing right after the preposition. Together, they create a Prepositional Phrase, along with other minor modifiers.

This Prepositional Phrase, as a unit, moves around the sentence just like other modifiers do. Prepositional Phrases can act as adjectives, modifying a noun: *The demon [**with** the three eyes].* They can be adverbs, modifying the verb: *The demon sat [**in** the chair].* Prepositional Phrases can even be adverbs that modify adjectives or other adverbs, creating nests in nests: *The demon [**with** the three eyes [**in** her face]]. The demon sat [**in** the chair [**with** the yellow pillow]].* When you find a preposition, look for its noun buddy nearby.

10. COORDINATING CONJUNCTIONS (cc)

As catchy as Schoolhouse Rock's "Conjunction Junction" song is, there's a little bit more to these connectors. There are two types: **coordinating conjunctions (cc)** and **subordinating conjunctions (sc)**. (Other grammarians sometimes refer to them as **coordinators** and **subordinators**. Everyone has a slightly different system of names, which is partly why grammar and physics are tough). Semantically, conjunctions are Function words that connect sentence elements together. Coordinating conjunctions link two *equivalent* nests placed adjacent to each other; subordinating conjunctions put subordinate nests *inside* the main nest. When you diagram sentences, a dotted line connects the two coordinated or subordinated elements together (noun to noun, verb to verb, etc.).

Syntactically—on the level of syntax or sentence order, looking forward to Lesson III—**coordinating conjunctions** link sentence elements that have the same syntactic role; that is, they must connect elements of the same level of hierarchy: independent clauses to independent clauses, dependent clauses to dependent clauses, phrases to phrases, and word classes to word classes. This **parallel structure** is important, and it's an issue I often spot when editing. Make sure you're using the same type of word or phrase on either side of coordinating conjunctions.

My eleventh-grade teacher taught us this handy acronym for coordinating conjunctions, FANBOYS:

For
And
Nor
But
Or
Yet
So

Correlative conjunctions are pairs of coordinating conjunctions connected to other words, making a pair that should not be interrupted with a comma in the sentence: *Choose **either** him **or** me. It is **not only** beautiful **but also** inexpensive. I wish **both** Bert **and** Ernie could be our neighbors.* Later, we'll use them to connect clauses to other clauses.

Binomial expressions are an easy way to see how parallel structure works even at the word class level. Verbs coordinate with verbs: *Let's <u>wait</u> **and** <u>see</u> what happens.* Connect nouns with nouns: *Do you want <u>this</u> **or** <u>that</u>?* Connect adverbs with adverbs: *I should have fought her <u>there</u> **and** <u>then</u>.* Connect adjectives: *Send your <u>best</u> **and** <u>brightest</u>. The movie's in <u>black</u> **and** <u>white</u>.*

Similarly, when you have a list of something, make sure all the elements in the list are the same word class (or same type of phrase or clause). Then use a coordinating conjunction before the last element in a series: *Tom, Dick, **and** Harry.* If you want to use a literary device called *polysyndeton* to make the list seem long and drawn out, use the coordinating conjunction between each word in the list: *Tom **and** Dick **and** Harry.* You can also use *asyndeton* and remove all connectors in the list to make it snappier: *Tom, Dick, Harry.* You can see how this literary device works at the sentence level in Table 4.1.

11. SUBORDINATING CONJUNCTIONS (sc)

Subordinating conjunctions skip right over phrases and introduce clauses known as **dependent clauses**, which are coming up in Lesson III. Dependent clauses can't stand alone but are embedded—nested—within the **independent clause**: *[I love physics [**because** it describes existence]].* You can't walk up to someone on the street and say, "Because it describes existence" and expect them to understand. You need the independent clause (*I love physics*) for the dependent clause to make sense. There are three major subclasses of subordinating conjunctions:

1. Most of the subordinating conjunctions introduce **adverbial clauses**. These dependent clauses modify a verb like adverbs do. They contribute details of time, place, reason, condition, and comparison to the independent clause: *[**If** you build it], they will come. I want to do it [**because** I can].* Subordinating conjunctions in this subclass include *after, because, if, since, although, while,* and others.

2. There are three subordinating conjunctions that introduce **degree clauses**: *as, than, that. The demon is not so intelligent [as his sister]. I love Max more [than sunflowers love sunshine]. I was so upset about Tom [that I couldn't think straight].*

3. There are three subordinating conjunctions that introduce **complement clauses**: *if, that,* and *whether. Longman* calls them **complementizers** because they complete the meaning of key words in the main clause: *I don't know [if it works]. I'm delighted [that you decided to join us].*

   Note: *Wh*-words and the relative pronoun *that*, which are not subordinating conjunctions, can also introduce dependent clauses as complementizers. Here's a subordinate clause, a relative clause, and a nested relative clause, in order: *[When Dyna comes on Sunday], I'll show her [what happens to demons [who cross me]].*

Complex subordinating conjunctions, like complex prepositions, can consist of more than one word: *as long as, as soon as, given (that), on condition (that), provided (that), except (that), so (that), as if, as though, even if, even though.* These are Strange quarks that can be simplified back down to Down quarks when possible.

12. Top Words (Insert Words)

We now review, again, the twelfth word class of Table 1.1, the stand-alone words science writers don't need. Interjections are interrup—

"*Ahem.*"

I spun around in my chair. Dyna was perched atop the office book-case behind me like a ghoulish Elf-on-the-Shelf.

"Just dropped in to see how you were doing, JZ. I see you're hard at work. I'll just pop into the kitchen and take the Scotch with me."

My dam burst. "Don't you have anything better to do?" I hissed, and for a moment, she appeared so startled she nearly fell off the shelf. "I had to tell Max that I'd left the back door open and the cat escaped. He was crushed. We spent the last two afternoons putting up Lost Cat signs all over the neighborhood. I felt horrible!"

"You shouldn't have lied, then. I'm not sure why parents ever lie. Just give bad news or awkward info tactfully and simply." She leapt to the floor and sauntered past me into the hallway while I chewed on her words with gritted teeth. A moment later, from the kitchen below my office, I heard the scraping of a chair and a clink of ice cubes. When she returned, she had a tumbler of ice in one hand and the whole Scotch bottle in the other.

"Don't let me interrupt," she said. "You're doing OK. It's a little dense, but then, most humans are. Remember what I said, don't discuss color charge. Lesson II is already too long and they don't care."

"Give Tom back now, and I'll finish the book as you ask," I pleaded. "Is he all right?"

"Oh, he's . . . fine, or not fine, depending on your frame of reference," she said airily. "I haven't checked. Carry on, carry on." She dissolved in a shower of sparks, this time marking my office floor with her signature starburst burn.

COLORFUL PHRASES

It is too late Tuesday night and I am too furious to fully discuss color charge in quarks and gluons, but I'm also too furious to listen to Dyna's advice, so here is a simplified summary of color charge in particle physics: we needed a system to describe how subatomic eggs balance, or add up to zero, inside a proton or neutron or pion nest—in the same way that "positive" and "negative" add up to zero for electric charge. But with two quarks in a meson and three in a baryon, how do you show a two-way *and* a three-way charge?

So physicists chose six "colors" to tell eggs from eggs and nests from nests, even though there is no color whatsoever at the subatomic level (hello, metaphor). They chose **Red**, **Green**, and **Blue** for quarks, and the complementary colors **Cyan, Magenta**, and **Yellow** for antiquarks (AntiRed, AntiGreen, and AntiBlue, respectively).

And they started metaphorically painting. Everywhere.

They said two quarks of the same color, like Blue and Blue, would repel each other like two negative electrons. In contrast, color and its complementary *anticolor*, like red and cyan (AntiRed) add up to zero, which is a balanced, colorless state—like how an equal push and pull cancel each other out.

Similarly, they said Red + Green + Blue—or their anticolor opposites, Cyan + Magenta + Yellow—also added up to a colorless state (in color theory, this is called the RGB additive model, where the three colors add up to white and the three anticolors add up to black). This arrangement created a balanced nest of quarks. The quarks within the nest exchange gluons (each of which has a different color/anticolor configuration) to keep a color charge balance, in the same way atoms add or give away electrons to balance electrical charge.

A proton, for example, with its Up + Up + Down quarks, will have one Red quark, one Green quark, and one Blue quark to remain stable, although it keeps up the balancing act by constantly exchanging gluons between the three colors.

For complicated matrix reasons, the eight independent gluon states have the worst names possible, like $(r\bar{b} + b\bar{r})/\sqrt{2}$, so I gave them more concrete color names and attached them to the eight kinds of phrases (see Table 2.2).

For Subatomic Writing, phrases in English must similarly balance, both internally and among all the nests. Multi-word phrases usually need a **head** plus whatever words, nested phrases, or even clauses **complete** or **modify** the head. Grouping words into their distinctive phrases will set the stage for syntax in Lesson III.

We have all the tools we need to make colorful phrases now—I can think of a few colorful phrases myself after Dyna's interruption. Let's run through the eight types of phrases, all of which require a head word in addition to other words and modifiers. I encourage my students to

TABLE 2.2

Gluon states and their corresponding English phrases, including composite colors (shown in the column labeled "A More Colorful Name"). The bold head words under "Organization" are essential to building the phrase; the rest are optional additions in the phrase.

GLUON	A MORE COLORFUL NAME	ENGLISH PHRASE	ABBR.	ORGANIZATION
Red/ AntiBlue	**Flush Orange**	Noun Phrase	NP	**head noun** + modifiers (mod) *the blue **demon***
Blue/ AntiGreen	**Electric Purple**	Verb Phrase	VP	**main verb** + aux + mod *will have been **trying***
Green/ AntiRed	**Spring Green**	Adjective Phrase	AdjP	**head adj** + mod ***faster** than I could blink*
Red/ AntiGreen	**Deep Pink**	Adverbial Phrase	AdvP	**head adv** + mod *far too **late***
Blue/ AntiRed	**Dodger Blue**	Prepositional Phrase	PP	**prep** + mod + **object of the preposition (OP)** + mod ***in** the **library** downstairs*
Green/ AntiBlue	**Bright Lime**	*VERBAL PHRASES (PIG)* — Participial Phrase	PartP	**participle (part)** + NP + mod ***watching** me with one eye*
		Infinitive Phrase	InfP	**infinitive (inf)** + NP + mod ***to read** books in the library*
		Gerund Phrase	GP	**gerund (ger)** + NP + mod ***waking** me up*
Brown/ AntiBrown	**Mocha Brown**	Appositive Phrase	App	**NP, rename NP,** ***Dyna,** my worst **nightmare,***
Gray/ AntiGray	**Gravel Gray**	Absolute Phrase	Abs	**NP + PartP** ***weather permitting***

memorize the questions they can ask themselves to identify these phrases. I also encourage any use of color to help you remember them.

Noun Phrase (NP)

A **Noun Phrase** answers the question, "**Who** or **what** does the action, receives the action, indirectly receives the action, or complements the action?" It consists of a **head noun** + any "adjective-like" modifiers: adjectives, determiners, Prepositional Phrases, and participial adjectives are the most common NP modifiers. Find the head noun, then identify modifiers that belong in the same NP nest.

A Noun Phrase often holds only one egg in its nest: *[chair]*. Or, the phrase can have a head noun plus modifiers—one big egg plus tinier eggs in the NP nest: *[the gray **chair**, burnt]*. This particular phrase has one head (*chair*) and three modifiers: a determiner (specifically, the definite article *the*), an adjective (specifically, the attributive adjective *gray*), and a participle acting as an adjective (specifically, the past participial adjective *burnt*).

Notice too that *the gray chair, burnt* is not yet a sentence; there is no main verb, so it's currently a fragment—as all phrases are.

You can even have a head egg in your NP nest plus another modifying phrase next to it, a nest holding its own eggs: ***chair** [in [the [**library**]]]*. *In the library* is a Prepositional Phrase, a nest unto itself, but it describes *which* chair we're talking about. And yes, "the library" is its own Noun Phrase inside the Prepositional Phrase. This is an example of **recursion**, that *Inception* of nests. Thankfully, our brains do most of this parsing work for us.

Pronouns can also stand as the head of Noun Phrases. The noun or noun-like entity that the pronoun refers back to is called the **antecedent**. One of the most frequent editing problems I come across is the **unclear antecedent**—which means the reader doesn't know what NP a pronoun is pointing back to. "It" and "This" are particularly problematic when referring to previous ideas or clauses. Whenever you use a pronoun, make sure it matches *the closest preceding noun* (or noun-like unit).

I once saw a meme (a cultural "emic unit") that said, "My therapist told me to write **letters** to the **people** you hate and then burn **them**. Did that, but now I don't know what to do with the letters." This meme is a great (but violent) example of how an unclear antecedent messes with the writer's meaning.

Verb Phrase (VP)

A **Verb Phrase** answers the question "What is the **action**?" A VP consists of a main verb (which could also be in participle form) + any auxiliary verbs. The main verb is the head and final verb in a Verb Phrase: *could have been **written***.

After finding the main verb and any auxiliaries, find modifiers that change the verb in some way. Adverbs, indirect objects, prepositions, infinitives, and dependent clauses most often modify a Verb Phrase. They might be part of their own Adverbial Phrase and are diagrammed underneath the main verb.

The **Long Verb Phrase**, also called the **predicate**, contains the verb and any actors or modifiers that receive its action: *The book [**will have been written** by Sunday at the latest to reclaim my cat from Dyna]*. If you're just beginning to diagram, find one Noun Phrase and one Long Verb Phrase in simple sentences: start to recognize the subject and predicate of the sentence.

As we covered in the Verbals section on PIGs, sometimes main verbs also take on **complementary infinitives** that complete their meaning: *I [ought [to report]] you.* NP + [VP [InfP]] + NP.

Adjective Phrase (AdjP)

Adjective Phrases, like adjectives, modify Noun Phrases. Adjectives don't often have their own nests; they are usually the modifiers in NP nests. Sometimes, however, an adjective acts as the head of an Adjective

Phrase, like *[too **eager** [to please]]* or *[more **vulgar** [than any creature I've ever seen]].* Adjective Phrases usually take on adverbs, Verbal Phrases like Infinitive Phrases, and dependent clauses like degree clauses.

If you think there's an AdjP nested inside an NP, remove one or two words so you can tell what is modifying what: *[an [excessively stormy] night].* If you take out *stormy*, you'll see that *an excessively night* doesn't make sense, so *excessively* must be an adverb modifying the adjective *stormy*. The article *an* still modifies the head noun (*night*) of the bigger NP nest.

Adverb Phrase (AdvP)

Adverbs act as the head of **Adverb Phrases**, like *quickly* in *more quickly.* Also called **adverbials**, they can modify verbs, adjectives, other adverbs, and verbals, along with their phrase-level equivalents.

Adverbs and Adverb Phrases answer the following questions:

Where does the action happen? *in the library*
When does the action happen? *late at night*
Why does it happen? *to irritate me*
How does it happen? **In what manner?** *[too **quickly** [for me [to react [properly]]]]*

Adverbs that answer these questions often use the *-ly* suffix, but not always.

Adverbial particles—yes, they're really called "adverbial particles"—are Function words that have a core meaning of motion. They are used to bulk up the multi-word phrasal verbs mentioned earlier: *go **away**, come **back**, put **on**.* Some people call them phrasal verbs or prepositional verbs, others call them verbs with adverbial particles. Students who are not native English speakers call them verbs from hell. I usually diagram adverbial particles that are essential to the meaning of the verb right next to the verb. If they can be deleted without ruining the meaning, then I diagram them underneath the verb as adverbs. If there's a noun after the adverbial particle, then be careful because that particle might actually be a preposition.

If you want to know if a word is an adverbial particle or a preposition, *Longman* has a test, and this test is called *particle movement*. See if you can move the particle before and after the noun or Noun Phrase. If you can move it and it still makes sense, then it's an adverbial particle: *I **picked up** the book. I **picked** the book **up**.* Since I can't delete *up* without ruining the meaning of the verb (*I picked the book* means something different than *I picked **up** the book*), I diagram the particle on the horizontal line right after the verb. If you can't move it, then it's a preposition, the beginning of a Prepositional Phrase.

Prepositional Phrase (PP)

Prepositional Phrases modify nouns, verbs, adjectives, adverbs, and even verbals. Depending on context, they can answer either the questions listed under "Adjectives" or the questions under "Adverb Phrases."

As a quick review, because many students stumble on Prepositional Phrases, you can identify the phrase by finding a preposition + the noun or noun-like unit that usually comes right after it. That constant companion of the preposition is the object of the preposition (OP), a glorified name for a nested noun or noun-like unit after the preposition.

The OP helps us identify which pronouns to use. We use **object pronouns** following a preposition: *[To **whom**] it may concern; Don't worry [about **Joe and him**]*. If you're not sure you're using the right OP form, do a deletion test. Take out the other person and see what sounds right: *Don't worry [about ~~Joe and~~ him]*. Educated speakers often "hypercorrect" and use a subject pronoun instead, but *Don't worry [about ~~Joe and~~ I]* does not sound right.

However, if an OP is a whole nested dependent clause itself inside the Prepositional Phrase, you may need to keep the subject pronoun: *I want to dance [with [**whoever likes the color blue**]]. Whomever* would be the wrong form in this case. It's this nasty nested stuff you have to watch out for.

Verbal Phrases: The Pig Strikes Back

In Minecraft, there's a point in the game where those terrible piglins can become zombified piglins. If you thought identifying one-word verbals was tough, it's even harder when verbals have their own monstrous nests filled recursively with even more monsters and their nests. The three kinds of **Verbal Phrases**—Participial, Infinitive, and Gerund Phrases—are verb-born constructions that start with a "simple" verbal—the vampiric participle, the primordial infinitive, or Gerald the friendly gerund. But then the PIG strikes: the three verbals enact their verb-like powers to trap other unsuspecting actors, actions, modifiers, and connectors into their own unholy phrase nests.

Some linguists call these phrases *ing*-**clauses**, *ed*-**clauses**, and *to*-**clauses** (or *-form* or other jargon), since the PIG can be both actor and action at the clause level (a clause needs both a subject and verb), but I think that's giving these monsters too much credit. In truth, PIG Phrases exist in the accursed realm between phrase and clause, just as PIG verbals exist between verbs and other word classes.

Participial Phrases (PartP). We've already met the participle in its half-verb incarnation, as a main verb with auxiliaries and modifiers: the Verb Phrase (*had been **writing**, is **written***). But when the participle is half verb, half adjective, the Participial Phrase consists of a **head participial adjective** + optional NP + optional modifiers. Such phrases look and act just as adjectives, but they have the same verb-like ability to take a direct object (to act upon a noun or noun-like unit). They can be modified by other words in their own right: *[**Eating** [a French fry] [as it flew]], the bird swooped overhead.* Participial Phrases also come in their past tense flavor and can appear after the noun they modify: *Jamie, [**destroyed** [by piglins]], wept in shame and frustration.*

Infinitive Phrase (InfP). We can identify an Infinitive Phrase by its **head infinitive** (*to* + base verb) + optional NP + optional modifiers. Here is the extended version of the previous infinitive examples:

- **complementary**: *Are you able [__to fight__ [the piglins] quickly]?*
- **nominal**: *I want [__to win__ [the game] [on the first try]].*
- **adjectival**: *Now is the right time [__to scream__ [your brains] out, loudly].*
- **adverbial**: *[__To win__ [shiny gold [from piglins]] [during the game]], you must barter and murder.*
- **bare**: *Help me [[__x__] survive [this lesson]].*

Gerund Phrase (GP). Gerund Phrases are close variations of Noun Phrases. They consist of a **head gerund** + optional NP + optional modifiers: *[[The competitive __swimming__] here, [among Marylanders]], is intense.* The modifiers can be determiners (*the*), adjectives (*competitive*), Prepositional Phrases (*among Marylanders*), and adverbs (*here*). They can create their own verb–object nest within the sentence: *[__Reading__ textbooks [at night]] is my favorite activity. I like [voraciously __eating__ food].*

Gerald, among his many activities, also loves to hang out with prepositions. He follows them around as the object of the preposition: *[After [__reading__]], I fell asleep.* But Gerald, being as friendly as he is, then drags other friends into his nest: *[After [__spending__ [[so much] time] apart]], we should hang out. After* is the preposition that kicks off the Prepositional Phrase, which is modifying the main verb as an adverb. *Spending* is the gerund and the head of its own internal Gerund Phrase. *So* is an adverb modifying *much*; *much* is the adjective head that modifies the NP head *time*; *apart* answers the question "Spending time how?" and is therefore an adverb modifying the gerund *spending*. *We* is the noun actor functioning as the subject of the independent clause, *should* is a modal auxiliary verb, and *hang out* is a phrasal verb, with *out* classified as an adverbial particle but diagrammed along with the verb. Whew.

Using word class abbreviations (see Table 1.1), this nasty sentence looks like this:

[Prep [gerund [[adv adj] n] adv]], n m.aux v.

In phrase abbreviations (see Table 2.2), the same sentence looks like this:

[PP [GP [[AdjP] NP]]], NP VP.

In clause abbreviations (see Table 3.1 in Lesson III), the sentence looks like this:

S V.

The sentence diagramming we'll review in Lesson III takes these still rather inaccessible groups of brackets and letters and graphs them on a coordinate axis. Exciting, I know.

We will flee from Gerald and the Verbal Phrases for now, but beware the PIGs and their penchant for nesting. Science writers fall too easily into their clutches.

Appositive Phrase (App)

An **Appositive Phrase** comes after a noun or noun-like unit and renames it, providing more information. The phrase should come right after what it renames: *Katie, [my sister], is a professional artist.* The Appositive Phrase *my sister* renames the proper noun *Katie*.

We thought we could escape the PIG, but we can't. Here's a more complicated appositive: *My goal, [[to get] my cat back], keeps me going.* We're renaming a noun, *goal*, which means we need another noun, noun phrase, or noun-like monstrosity like an Infinitive Phrase. The two commas on either side of the Appositive Phrase are a major hint that an appositive is at work.

If the phrase doesn't have commas, as in *My sister [Katie]*, then it's considered "restrictive," meaning that if there are no commas, I'm communicating to the reader that I have more than one sister (which

I do!), but right now I'm referring to the sister named Katie. Writers make this mistake all the time when writing about their spouse, which I think is funny. *My wife Jessica is great* is stating three things: Jessica is a wife, she is great, and she's not the speaker's only wife. Your wife, Jessica, would like you to remember how to use commas around appositives.

Absolute Phrase (Abs)

A variation of the Prepositional Phrase is the **Absolute Phrase**, which is a head noun + any modifiers: *[**Lips blue**], she reached for her coat.* Absolute phrases usually come at the beginning or end of the sentence. They're loosely connected to some part of the sentence, if at all. We approach them with caution.

And just when you thought we were done with the chapter, the PIG returns for one final battle. All three Verbal Phrases make a final strike as they riotously make their nests inside an Absolute Phrase: *They laughed, their jeering growing louder in the night to mock their prey: you.*

Can you pick out what's happening? Which one is the blood-sucking participle (in which form, as half verb or as half verb, half adjective?), where is the primordial infinitive and whom has she attached herself to, what happened to Gerald? Where are the nests and the nests in nests in nests, with the other sad, sticky words and phrases trapped among them? Here's a hint: *They laughed, [their **jeering** [growing louder [in the night][to mock [their prey: [you]]]]].*

Perhaps the answers will come to you in the night.

Are we sick of monsters and PIGs? Of birds and sticky nests? I am. But they've served their purpose. Gluons not only glue quarks into composite particles; they also glue these larger composite particles to each other, just as phrase rules glue words into larger and larger units. This is where we—at last—transition to syntax.

EXERCISE 1

Take a favorite sentence from a favorite author and divide it into its smallest parts using Easy, Medium, or Advanced grammar jargon. Identify the word class of each word with abbreviations from Table 1.1. Then, group the words: find their phrase units and use the abbreviations from Table 2.2. If you have access to colors like Flush Orange and Spring Green, you can have your own round of subatomic paintball by marking phrases with color. Optional: Post your colorfully glued phrases to Twitter.

EXERCISE 2

Take a paragraph of your own writing. Parse and tag each word class and each phrase (using Easy, Medium, or Advanced grammar jargon). Are you missing any modifiers? Objects of the preposition? Are certain types of nests missing from your repertoire? If you're having trouble identifying the word class of a word, I highly suggest finding the word on *Merriam-Webster*'s online dictionary (www.m-w.com) and sorting through the definitions to find which word class fits your specific use of the word.

VISUAL SYNTAX

Clause Level

And the words slide into the slots ordained by syntax, and glitter as with atmospheric dust with those impurities which we call meaning.

—Anthony Burgess, *Enderby Outside* (1968)

WE'VE NOW USED THE VIBRATIONS of sound and movement to amass letters into meaningful words, with the help of the literary Higgs mechanism. We've also stuck words into eight kinds of balanced phrases with the help of eight literary gluons, and now we're going to fuse those phrases into coherent clauses with the help of our third exchange particle, the photon.

The **photon** is a bit of light, a zero-mass particle that carries the **electromagnetic force**. (The wave nature of light is essential too, but we'll stick with particles for now.) James Clerk Maxwell, building on earlier work from Michael Faraday and Hans Oersted, mathematically fused the laws of electricity and the laws of magnetism in 1873, giving the made-up combo name *electromagnetism* to a real phenomenon.

Consider the simple act of flipping on a light switch. The stream of photons from a light source—including the blue LED from my bedside phone charger—shows the electromagnetic force at play. Photons scatter here and there and everywhere in the universe, illuminating the

103

darkness, revealing the mysteries of matter—or at least the demon in my bedroom. Almost all information from the cosmos comes to us in the form of light: when our eyes gather light, we glean—extract—a surprising amount of information about an object. Light sensing is a technique ancient organisms developed at the beginning of the Cambrian period, about 541 million years ago. The organisms that developed eyes, specifically eyes that detect photons scattering off predators, lived longer. Detection meant survival.

The very characteristics of light scattering back to us from an object can reveal information. The "cursed" Hope Diamond at the Smithsonian Museum of Natural History, for example, has a lusty steel-blue color and is about the size and shape of a pigeon egg (you thought we were done with eggs). The diamond's crystalline structure, like all diamonds, is made from carbon atoms—one carbon atom is six protons, six neutrons, and six electrons, a fact I suspect Dyna appreciates. But the blue-wavelength photons reflecting off the diamond back to our eyes reveal that the organized carbon is also infused with trace elements of boron, an atom with five protons in it. Light is information.

Matter can also generate photons itself—through fusion, as our Sun and stars do. When I taught astronomy in Nicaragua, I loved teaching my students how stars fuse protons together to generate light, energy, and the heavier atomic elements we benefit from here on Earth. Here is the simplified story of fusion: Our faithful proton from Lesson II— two Ups and one Down, glued together into one composite particle— makes up the nucleus of the **hydrogen** atom, the first element on the periodic table. The proton has a positive charge, unlike the neutral charge of neutrons or the negative charge of electrons. Our one positive proton zips by other protons inside a star; like two north poles of a magnet, two protons usually repel each other, like two women dressed the same at a party. However, in the core of stars, at crushing pressures and temperatures, the protons hit each other so hard they collide and stick together—fuse—to form a heavier nucleus. The two women at the party become best friends and start to go everywhere together.

Sometimes one of the protons, in the process, changes from a proton to a neutron. If this happens, these two composite particles have just fused to form **deuterium**, an isotope (a variation) of hydrogen whose atomic nucleus holds one proton and one neutron ($\mathbf{p}+\mathbf{n}$). A neutrino and other bits are also produced along the way.

We've already seen how gluons exchange the **strong force** to keeps quarks together, but the *residual strong force*—lingering gluon action—also takes on the responsibility of beating back the electromagnetic force just enough to keep the proton and neutron together in a stable nucleus.

I'll skip the other fascinating details, but this is the essential pattern to expand the circle of proton and neutron friends: Hydrogen (one proton) fuses into deuterium (one proton, one neutron). Deuterium, in turn, fuses into **helium-3** (two protons, one neutron), which then fuses into **helium-4** (two protons, two neutrons). The whole process is called the *Proton–Proton Chain*. This is fusion, creating light, generating photons, allowing humanity to gather information and survive. (If you're still keeping track of quarks, this gives us a total of twelve Up and Down quarks: $\mathbf{uud}+\mathbf{uud}+\mathbf{ddu}+\mathbf{ddu}$.)

Writers fuse certain phrases together in a similar chain-like fusion process. We start with a literary proton, a **fragment**—a phrase or just one word. A fragment isn't a complete thought, but in the right context, when accelerated by the field of words around them, they are mighty powerful for such a small particle. The Oh-My-God particle, the most energetic of all detected cosmic rays, was captured in 1991 by our Fly's Eye detector at the University of Utah. It was most likely a single proton.

Now we'll take our lone fragment and fuse it to another fragment. By fusing a Noun Phrase (or any noun-like unit) to a Verb Phrase ($NP+VP$), we create a **clause**. This new Subject + Verb unit ($S+V$) stays glued together using the literary residual strong force, which are the subject-verb agreement rules mentioned a little later in this lesson. Like a star fusing two hydrogen nuclei into one deuterium nucleus, you've just formed literary deuterium, the atomic backbone of any sentence, creating illumination and energy for the reader in the process.

A complete sentence must have at least one **independent clause**, an S + V entity that forms a complete thought: *Dyna arrives.* From there, we fuse more particles to deuterium according to syntax rules. In so doing, we create complex literary nuclei like helium-3 and helium-4, completing the Proton–Proton Chain as stars do. At any time, a paragraph has a healthy mix of different patterns and their variations (many of which are shown in Table 4.1). Using only one pattern bores the reader.

In addition to one or more independent clauses, a sentence may also contain optional **dependent clauses**, less important S + V clauses. Sometimes a dependent clause has a subordinating conjunction (sc) in front of it: *We talked [**until** the sun came up]. [**When** you get there], call me.* If you deleted the dependent clause, the independent clause would still make sense by itself: *We talked. Call me.* Other times, the dependent clause slides into a section of the independent clause: *I knew [**that** she could do it]. [**What** we want is] not always [**what** we need].* If you walk up to someone on the street and speak in dependent clauses only, the stranger is unlikely to understand you: *Until the sun goes down. When you get there. That she could do it. What we want. What we need.*

Whether we are generating science information ourselves as the Sun generates light—or reflecting information from others as the Moon reflects photons from the Sun—our writing should illuminate the reader in some way. When we fuse words, phrases, and clauses into chains of coherent syntax, we are increasing the probability that our readers will have that light-bulb moment.

On the other hand, if our syntax sucks, if the arrangement of our phrases and clauses doesn't reflect or generate light, then the words will provide no illumination, no information. The sentence becomes a dead star collapsing in on itself, so dense that no light—no information—can escape. I like black holes, but I don't like black-hole writing.

As one of my favorite authors, Ray Bradbury, says, "Let the world burn through you. Throw the prism light, white hot, on the paper." In the rest of this lesson, we will introduce the history and secrets of diagramming particles—in both science and writing. We'll define the names of the various "slots" in English word order—**syntactic roles**—and talk about how these slots are ordered in a sentence, which we call **syntax**. Finally, we'll discuss the types of independent and dependent clauses and best practices for concision at the clause level. I've added many of my favorite illuminating quotes in the exercises at the end so you can practice visualizing syntax.

FEYNMAN DIAGRAMS AND SENTENCE DIAGRAMS

Theoretical physicist Richard Feynman was disappointed and even depressed, in 1948, when no one at the Pocono Conference (the second of three important postwar conferences held in Pennsylvania), liked the doodly diagrams he introduced to explain quantum physics. As physics historian David Kaiser described, "He suffered frequent interruptions from the likes of Niels Bohr, Paul Dirac and Edward Teller, [who each] pressed Feynman on how his new doodles fit in with the established principle of quantum physics." Thanks to Feynman's dedicated assistant, Freeman Dyson, the doodles eventually caught on, improved, and helped physicists actively solve quantum physics problems. The diagrams turned hairy equations into simple spatial pictures: the drawing showed which particles affect other particles in specific ways, plus the when and how of the interaction.

Although you'll find variations, Feynman diagrams usually have a vertical "Time" axis and a horizontal "Space" axis. They show, with various types of lines, what happens when elementary particles collide. We draw quarks and leptons as **straight lines** with arrows, representing the "probability amplitude" of a particle going from one place to another, and it can go back and forth in time (if the arrow points downward, it's an antimatter particle). A photon is a **wavy line**. A gluon line looks like a **spring**. And finally, **dashed lines** are Z and W bo-sons. Where these lines meet up are called **vertices**.

That's about it.

For example, Figure 3.1 shows the strong residual force holding a proton and a neutron together with a gluon: the incredible deuterium. Dalton's ball-like particles have been recast as lines in Feynman dia-grams, and we'll eventually do the same in sentence diagramming. For more on forming Feynman diagrams, I recommend the YouTube video "How to Read Feynman Diagrams" by Domain of Science as a good, simple introduction.

In 1847, a hundred years before Feynman introduced his diagrams, a schoolmaster named Stephen W. Clark, in Homer, New York, created a

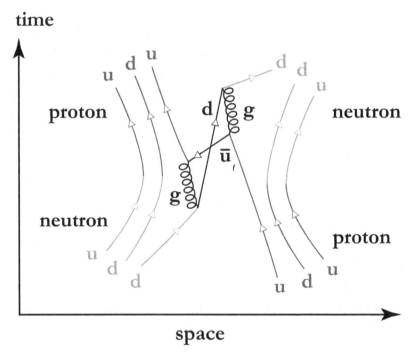

FIGURE 3.1 A deuterium nucleus—one proton and one neutron—being held together by the residual strong force (**d ū** is a negatively charged pion).
Source: Jamie Zvirzdin.

visual system that grouped words according to their syntactic functions—how they related to each other—in a sentence. *"Grammar,"* Clark stated in his book *A Practical Grammar*, "is the Science of Language." He used linking bubbles—similar to John Dalton's early chemistry models—to show how the various particles of language collided with each other and even nested inside each other. I like how he divides all those word classes and phrases we learned in Lesson II into two simple categories: *Principle Parts* and *Adjuncts*. This is how I see sentences. We have the important "core" of a sentence—what *Longman* calls **valency patterns**—and the less important modifiers of those core parts. We'll come back to valency patterns in a moment.

"Let then the use of Diagrams," Clark declares, "be adopted in the analyses of sentences . . . that an abstract truth is made tangible, the eye permitted to assist the mind, the memory is relieved that the judgment may have a full charter of all the mental powers."

Kitty Burns Florey, author of the fantastic and funny *Sister Bernadette's Barking Dog: The Quirky History and Lost Art of Diagramming Sentences*, said in a *New York Times* op-ed that Clark's balloons were more like a family of hot-dog rolls. (See Figures 3.2 and 3.3.) However, I kind of like them, and we have Clark to thank for doing the same thing for English language studies that Feynman did for subatomic physics, an easier spatial representation so we don't have to endlessly parse everything with nested brackets.

In 1877, Alonzo Reed and Brainerd Kellogg, two professors from the Brooklyn Polytechnic Institute, turned Clark's balloons into lines. While others have improved on the Reed-Kellogg system since then, this is the system my first-grade teacher, Miss Labrum, taught me, and I'm deeply grateful to her for doing so. I then learned sentence diagramming at a deeper level in third grade, and a deeper level still in seventh grade.

Then sentence diagramming disappeared. I had great English teachers after that, but none of them used sentence diagrams. My four younger siblings never received the same visual instruction of language

Exercises.

" *Truth, crushed to earth, will rise again.*"

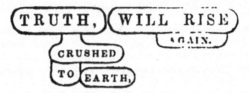

ANALYSIS.

Principal Parts....	{ Truth............Subject,	} Simple Sentence,
	{ Will rise..........Predicate,	} Intransitive.
Adjuncts.........	{ Crushed to earth.........Adjunct of " truth."
	{ Again.........................Adjunct of " will rise."

FIGURE 3.2 Original sentence diagram (in "particle" form).
The sentence diagrammed is *Truth, crushed to earth, will rise again.*
For us, this is an S + V construction, with *Truth* as the Subject,
will rise as the Verb, *crushed to earth* as a past participle phrase
(the past participial adjective is itself modified by
a prepositional phrase, *to earth*) and *again* as an adverb.
Source: Stephen W. Clark, *A Practical Grammar* [. . .]. New York: A. S. Barnes, 1847.

that I did. Some studies said diagrams weren't helpful for some students, but for me diagrams showed the fundamental relationships of language: this spatial mental model helps me catch errors as I write and edit. I don't think I'm special in this regard—I believe anyone can learn to diagram.

Fortunately, Kitty Burns Florey and many others (Judith Coats, *A Sentence Diagramming Primer*; Eugene R. Moutoux, *Drawing Sentences: A Guide to Diagramming*; Elizabeth O'Brien, *Grammar Revolution*) have brought Reed-Kellogg sentence diagramming back, and I'm doing my part to pass on the practice. Earlier this year, when Max was struggling to identify parts of a sentence, I pulled out some paper and drew a bunch of solid and dotted lines. That's when it clicked for him. I've witnessed the same moment of illumination for many of my students when we diagram sentences together. In fact, I have made a few changes to the system, which I hope will make sentence diagramming even more useful. You can learn more about Subatomic Writing diagrams in the appendix.

" Time slept on flowers and lent his glass to hope."

SIGOURNEY.

(23.)

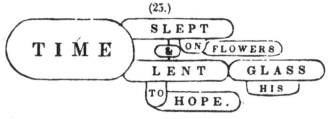

A mixed sentence.—Def. 30, *b.*

ANALYSIS.

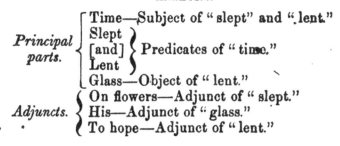

FIGURE 3.3 Another particle diagram: *Time slept on flowers and lent his glass to hope.* This is called a compound predicate sentence, where one subject fuses with two verbs. Notice that because a comma should never come between a subject and its verb, there is no comma after *flowers* before the second verb *lent.* You can help readers know that the subject is fused to both verbs by not putting a comma after *flowers.* (For more on the "compound predicate problem," see *The Chicago Manual of Style*, 17th edition, 6.22–6.23.)

Source: Stephen W. Clark, *A Practical Grammar* [. . .]. New York: A. S. Barnes, 1847.

If you want to craft a memorable sentence, study your favorite sentences from literature. What are great authors doing on a syntactic level? How do they order words, phrases, and clauses? What patterns are they using? The rest of this lesson will give you the vocabulary you need to identify what is happening in effective, beautiful writing, which you can then apply to your own fusion process.

A note on tree diagrams: Linguists prefer a separate diagramming system called *tree diagrams*, but I find these diagrams less visually appealing. They stay at the level of phrases, and while that's beneficial for many people, for me it's easier to spot valency problems and concision issues in Reed-Kellogg diagrams. When I ask my students to diagram sentences, however, I let them choose their preferred system: linguistic trees, Reed-Kellogg lines, Clark's bubbles, full Subatomic Writing diagrams, or a system of their own invention.

SENTENCE DIAGRAMS BY EUGENE R. MOUTOUX

Rather than replicate already great work—and besides, I only have until Sunday—I have summarized the basics of diagramming in the "Diagramming Valency Patterns and Modifiers" section of the appendix. If this summary isn't enough for you, head over to http://www.german-latin-english.com/diagrams.htm. Eugene Moutoux has already done an incredible and thorough job teaching the basics of sentence diagramming. I mostly use his system of Reed-Kellogg diagramming, and he has hundreds of examples. He also diagrams famous long sentences from literature that could not possibly fit on one page of this little book. I recommend you try to diagram the sentence yourself first, then check your diagram against his.

What I particularly love about Moutoux's approach is that he's taught both Latin and German, and because he knows these languages, he is thoroughly familiar with the roots of English. He taught at four universities and three high schools before he retired in 2004 and began building his online diagram empire. Like me, he learned how to diagram as a child from his elementary school teacher.

Another great website is Elizabeth O'Brien's *Grammar Revolution*, particularly the index, which contains examples of each item we've discussed up to this point: https://www.english-grammar-revolution.com/diagramming-sentences.html. You could learn the basics of diagramming in about ten minutes with these resources. Depending on what you already know (and how closely you've followed along!), you can learn advanced diagramming in thirty minutes to an hour. This is time invested in yourself, and you can test your investment using the exercises at the end of this lesson.

HOW ENGLISH SYNTAX
IS AND ISN'T LIKE ATOMIC NUCLEI

In English, word order matters, and this word order is called **syntax**, from the Greek *syn* ("together") and *taxis* ("an ordering"). A simple sentence like *Jamie hates the piglin* is different than *The piglin hates Jamie*. **Syntactic roles** determine who is the actor, what action the actor is doing, what or whom the action is affecting, and so forth. We will slide individual words, phrases, or clauses into these syntactic slots, many of which we're familiar with because of Lesson II.

Take a look at Table 3.1. We can learn English syntax relatively well by comparing the Proton–Proton Chain to major valency patterns . . . if we're a little flexible with our quark metaphor now. Technically, you can still force the subject of your sentence (S) to always have exactly three common Lexical or Function words—two Up words and one Down word in a Noun Phrase, as we've previously defined literary quarks (see Tables 1.1 and 2.2). You can still force the verb or Verb Phrase of a sentence (V) to always have exactly three words: two Down words and one Up word. Here are some examples:

p:	[Not you again!]	[u d u]
p + n:	[My poor cat] [has been nabbed].	[d u u] [d d u]
p + n + p:	[Max, her son], [will soon be] [taller than Andrew].	[u d u] [d u d] [u d u]

While it's fun to form strict three-word protons and neutrons (and you might want to, as an exercise), this rigidity of expression is neither advisable nor necessary. There's no way around it: the literary universe diverges from the physical universe when we move from phrases to clauses. We don't always need three-word ("baryonic") phrases to slot into one syntax role, especially when one word will do.

So I will channel James Clerk Maxwell here and say that particles of matter are *like* particles of language, as particles are *like* billiard balls, but of course it's not the same. Language is (thankfully) more flexible than subatomic and atomic structures. That said, check out the valency patterns and the examples for each in Table 3.1. I've also added **tritium**, another isotope of hydrogen, for one of the less common copular patterns. You can see each of these patterns diagrammed in Figure A.2 of the appendix.

I have one more important note about Table 3.1 before we define syntactic roles and types of clauses. Consider the simple intransitive pattern ***Dyna arrived***. It's a full sentence, a complete thought, an independent clause that can stand on its own two legs. It's a noun and a verb, a one-word Noun Phrase plus a one-word Verb Phrase, a subject and predicate. All the other adjective or adverb words (modifiers) that we could add to this p + n valency pattern do nothing more than modify the original subject and verb. For example:

> To my intense and everlasting dismay, **Dyna**, a vulgar blue
> demon from ancient Greece (who calls herself, also to my
> great dismay, Maxwell's mother's demon), **arrived** suddenly
> in the middle of the night to nab my cat and force me to write
> a strange book on how particles of language are like particles of
> matter, a metaphor that has already caused me a great deal
> of trouble but also, weirdly, joy—particularly in Lesson III.

The head noun that acts as the subject of the sentence is still ***Dyna***, and the main verb is still ***arrived***. Finding the subject + verb core of any sentence is the first challenge of the newly initiated diagrammer. You can see how all other modifiers—whether they are individual words, multi-word phrases, nested PIGs, or dependent clauses—can

TABLE 3.1

Major **valency patterns** of English syntax, matched up with the simplest nuclei in the universe. Minus the fragment, each of these patterns creates one independent clause.

ELEMENT	NUCLEUS	:	VALENCY PATTERN	ABBR.
Hydrogen H	p	:	Subject (fragment: *The blue demon.*)	S
Deuterium ²H	p n	:	Subject + Verb (intransitive: *Dyna arrived.*)	S V
Tritium ³H	p n n	:	Subject + Copular Verb + Obligatory Adverbial (less common copular: *Dyna was in my library.*)	S V A
Helium-3 ³He	p n p	:	Subject + Copular Verb + Subject Predicate (copular: *Dyna was a demon. She felt upset.*)	S V SP
			Subject + Verb + Direct Object (monotransitive: *Dyna nabbed my cat.*)	S V DO
Helium-4 ⁴He	p n p n	:	Subj. + Verb + Indirect Object + Direct Object (ditransitive: *I gave her the glass.*)	S V IO DO
			Subj. + Verb + Direct Obj. + Obj. Complement (complex transitive: *She called herself Dyna. Physics makes me happy.*)	S V DO OC
			Subj. + Verb + Direct Obj. + Oblig. Adverbial (complex transitive: *I put the book on the shelf.*)	S V DO A

cluster, expand, and nest before, between, and after the main subject and the main verb. These modifiers can also make the reader wander far, far away from the core of the original statement. Strong noun subjects, with strong verbs nearby, is the name of the syntax game for science writers. If you want to improve your writing, check that the nucleus of each sentence shines with bright Lexical words.

SYNTACTIC ROLES DEFINED

As we rise up the ranks of grammar units (word, phrase, clause, sentence, super-sentence, and paragraph levels), each level has its own name for things—too many names, just like in particle physics. Linguists and physicists alike argue about exact definitions and naming conventions, which makes it worse. But take heart: as you add two rich layers of jargon to your English cake, your cake becomes even more delicious. The literary and physical universes are even more astonishing and radiant when you can name and follow their inner workings.

At the syntax level, we continue to plug earlier units into a linear word order to communicate who is actually doing what to whom. Word classes told us the different actors and actions, modifiers and connectors, but syntax puts the story in motion. The Subatomic Writing diagrams and explanations in the appendix can help you see the relationship between levels.

Subject (S)

Your subject is your key actor, the main star of the sentence, plus any of its modifiers. Structurally, the subject is a noun, a Noun Phrase, or a noun-like word, phrase, or dependent clause. The subject normally comes before the verb, except in clauses with **inversion**, which is a nice changeup but will Yoda-fy your sentence if you're not careful.

The subject determines the **person** (first, second, or third) and **number** (singular or plural) that must match the person and number of the verb. This match is called **subject-verb agreement**.

If you've learned a language, you've probably seen conjugation charts, which are easy ways to see the difference between person and number as you're matching up subjects and verbs. Here's a list of all the personal pronouns screaming for ice cream:

First, second, and third person, singular: **I** scream, **you** scream, **he/she/it** screams.
First, second, and third person, plural: **We** scream, **you (all)** scream, **they** scream.

(Note that *they scream* can also be used in a singular sense, when the gender of one single person is not important or if *they* is their preferred pronoun. Again, if Shakespeare and Jane Austen did it, we can too.)

You might be surprised to hear that subject-verb agreement is an editing issue I still address with my clients, my graduate students, and my own editors. Agreement gets much trickier with complicated Noun Phrases, collective nouns, indefinite pronouns, relative pronouns, false attraction to predicate nouns, and misleading connectives. (See *The Chicago Manual of Style,* 17th Edition, 5.138–5.142 for examples of these.)

Sometimes we use the dummy pronoun *it* to fill the place of the subject, even if it has no substance itself and doesn't refer to any antecedent: *It surprised me to hear she was in town. It's warm in here.* Same with the existential *there*: *There is no place like home. There she goes again.* We need these constructions, but we shouldn't rely on them too often in writing.

Since the Subject + Verb is the heart and soul of your sentence, make sure these roles usually stay close to each other in a sentence. If they're too far away, interrupted by too many intervening modifiers and lesser clauses, the link between them is broken for the reader.

Verb (V)

The syntactic role of the verb, with the help of any of its lesser auxiliaries and modifiers in a Verb Phrase, is to carry out the action of the actor.

This action controls other "Principle Parts" (as Clark categorized them) of the core syntax pattern. Mercifully, we already covered verbs and Verb Phrases in Lesson II. We just have a few more terms to add.

Copular verbs, also known as **linking verbs**, identify an equal relationship between the subject (S) and the subject predicate (SP). (Some syntacticians—yes, this is a profession—call the subject predicate "subject complement" instead. I've chosen what I feel are the least confusing terms, particularly since we use abbreviations while diagramming.) A copular verb is like an equal sign because it helps the subject and the subject predicate hook up. In the following examples, the copular verbs are in boldface, and the two things—the nouns or adjectives—equated are underlined: *The cheese **has gone** moldy. It **is** I. She's a physicist. The young man **felt** bad* about it (not *badly*, which is an adverb and can't describe the noun *man*). Other copular verbs include *appear, be, become, feel, get, grow, go, keep, look, prove, remain, seem, smell, sound, stay, taste,* and *turn*. I picture these verbs with a little equal sign above them when I diagram sentences: Cheese=moldy. It (a dummy pronoun)=I. She=physicist. Man=bad (in the sense that *bad* is an emotional state describing how the young man feels, equating a noun with a quality or characteristic). At the end of this lesson, in Figure 3.4, you can see a Subatomic Writing diagram of a copular verb construction.

A **transitive verb** requires a direct object (DO) to complete its meaning: *Julia **makes** glassware. Julia makes*, on its own, isn't a complete sentence. The noun *glassware*, underlined as the direct object, completes the sentence. This is what we mean when we say, "the verb takes an object" or "the Verbal Phrase takes an object." It means that someone or something—the direct object—needs to come after the verb or verbal and receive the action.

An **intransitive verb** cannot take a direct object: *Stay [with me]*. Intransitive verbs stand on their own or can take other Adverb Phrases, including the Prepositional Phrase *with me*, but these kinds of verbs don't take Noun Phrases.

Some verbs are both intransitive and transitive: *I **write***. *I **write** essays*. Verbs that require two objects—a direct object and an indirect object—are called **ditransitive verbs**: *I **gave** her the book*. And yes, there is such a thing as a **tritransitive verb**, which takes a direct object, an indirect object, and a Prepositional Phrase or dependent clause: *I **bet** you a dollar [that he won't show]*.

Objects (DO and IO)

An object is a Noun Phrase that normally follows the verb. It can only occur with transitive verbs. There are two kinds of objects: direct and indirect. I'll boldface verbs and underline objects.

A **direct object (DO)** usually follows the verb, except when an indirect object butts in. The role of the DO is to declare which word, phrase, or dependent clause is affected by the action or process of the verb: *The demon **wanted** a drink*. Direct objects do not follow copular verbs—subject predicates (SP) do. Direct objects are always acted upon in some way; they are not equated with the subject.

Don't forget Gerald the gerund from Lesson II: sometimes a phrase headed by a gerund can act as the direct object: *We **should show** [a greater understanding for the complexity of human interaction]*. While I've put the whole direct object—in this case, a Gerund Phrase—in brackets, the head word of the DO is the gerund *understanding*. *We **should show** understanding* is the "unmodified" or core version of the S + V + DO valency pattern.

The same is true for Infinitive Phrases: *I **want** [to read too many books]*. Using a variable, like *x*, can help you pick out the direct object: I want what? I want *x*. What is *x*? *To read too many books*. Thus *x*, the Infinitive Phrase, is the direct object of this particular sentence.

Dummy pronouns can also be direct objects: ***Take** it easy, man*.

An **indirect object (IO)** is in clauses with verbs like *give* and *tell*. In the sentence ***Tell** me a story*, our subject is an invisible *you*, dropped to make a command: *[You] **Tell** me a story*. The verb is *tell*, the indirect object is *me*, and the direct object is *a story*. Ask: Who is the story being told to? To *me*. Indirect objects can often be identified by asking: To whom? Or for whom?

If you make the indirect object the subject, you can often change the voice from active to passive:

> I **will show** <u>you</u> *a new life.*
> <u>You</u> **will be shown** *a new life by me.*

Not that you'd want to do that, however. It's fine to use passive voice, but readers do prefer strong agents and strong actions.

Subject Predicate and Object Complement (SP and OC)

English has two syntactic roles whose job it is to modify the preceding noun. Skipping over the sordid details of *predicatives* and *complements*, I call these two roles the **subject predicate (SP)** and **the object complement (OC).** I find these names less confusing, and their unique abbreviations allow us to distinguish from earlier abbreviations defined in Lesson II, which is useful for diagramming. You can call them what you want. The SP and OC are usually Adjective Phrases, Noun Phrases, Verbal Phrases, or embedded dependent clauses.

The **subject predicate (SP)** is a Noun Phrase ("predicate noun") or Adjective Phrase ("predicate adjective") that renames the subject. Since the PIG from Lesson II can function as either a noun or an adjective, PIGs can be subject predicates too.

Subject predicates come right after copular verbs, the equal-sign-like verb. Here are some examples (subject underlined, subject predicate in bold):

Subject is equated with a Noun Phrase: <u>My teacher</u> is **a grumpy old witch**. Teacher = witch.

Subject is equated with an Adjective Phrase: *Her eyes appear* **very blue**. Eyes = blue.

Subject is equated with a Participial Phrase: *The experience had been* **so breathtaking**. Experience = breathtaking.

Object Complement (OC): This clause element completes—complements—the direct object. OCs come right after the direct object, and the main verb must be a complex transitive verb like *make, find, consider,* or *name.* Here are some examples (direct object underlined, object complement in bold):

Direct object is complemented by an adjective: *I can't get <u>this pickle jar</u>* **open**.

Direct object is complemented by a Noun Phrase: *I consider <u>my new computer</u>* **a miracle**.

Direct object is complemented by an Infinitive Phrase: *I found <u>my demonic muse</u>* **to be unexpectedly worthless**.

Obligatory Adverbial (A) and Optional Adverbial (OA)

Adverbials are simply adverbs by another name at the syntax level of language. There are two kinds of adverbials, **obligatory adverbials** and **optional adverbials**. An adverbial slot in a clause answers questions such as *when, where, why,* and *how.* An adverbial can be a single-word adverb, a multi-word Adverb Phrase, a Prepositional Phrase, or other adverb-like constructions.

Obligatory adverbials (A): Some verbs demand an adverbial to complete their structure and meaning. If you take out the adverbial, the sentence is no longer complete. This kind of adverbial usually denotes place or direction, but it can also denote time or manner. Here are examples of these less common S + V + A and S + V + DO + A valency patterns (subject in brackets, direct object underlined, adverbial in bold):

(S V A) *[Your dinner] is **in the fridge**.* (It doesn't make sense to say *Your dinner is.*)
(S V A) *[The meeting] lasted **late into the night**.*
(S V A) *[His ailing grandma] seemed **in bad shape**.*

(S V DO A) *[We] treated* <u>them</u> ***badly, very badly***.
(S V DO A) *[She] placed* <u>the telescope</u> ***on its mount in the observatory***.

You'll notice that the first set of examples (S V A) is very similar to the copular valency pattern S V SP. Likewise, the second set of examples (S V DO A) is similar to the valency pattern S V DO OC.

Optional Adverbials (OA): You can safely remove these modifiers and not damage the fundamental structure of the sentence. Writers are often advised to remove as many adverbials as they can because they clutter the core valency pattern. Optional adverbials can occur anywhere in the sentence and are loosely attached. They provide information like place, time, manner, extent, and attitude (*when, where, why,* and *how*). Examples:

*I **merely** decided to pop in **today**.*
*She was here, **with her mother, just a few minutes ago**.*
(Notice that *here* is an obligatory adverbial.)
*Writers will **thus** benefit **immensely from the intense study of grammar**.*

Peripheral Elements

Like the oddball phrase elements in Lesson II, like existential *there* and *wh*-words, we have clause elements that don't fit into these other categories. Briefly, they are:

- **Conjunctions** at the beginning of sentences: ***And*** *that's why it happened the way it did.* (This is totally fine and legal; just don't overdo it.)
- **Parentheticals** set off from the clause by parentheses: *I saw him **(or thought I did)** as I left the station.*
- **Prefaces**, which are sort of like misplaced appositives: ***That lady,*** *she's crazy.*
- **Tags**: *We showed them, **didn't we**?*
- **Inserts**, usually in dialogue: ***Hello***, *what is going on here? You know the homework's due today, **right**?*
- **Vocatives**, identifying the person or people addressed: *Come on, **you cowards**! **Reader**, I married him.*

Next, we'll cover a few more things about different kinds of sentences, and then we'll move to visualizing sentences with diagrams.

MOOD, INDEPENDENT CLAUSES, AND DEPENDENT CLAUSES

Let's look at the final big-picture level of clauses that create a sentence (although it won't be a full sentence without punctuation, which we'll cover in Lesson IV).

Mood

Sentences can change verb form and syntax to reveal how the speaker feels about the situation—to relay their intent or how certain they are of their information. Does the speaker want to state a fact? Ask why their facts weren't followed? Demand retribution in blood? Express the wish that they hadn't demanded retribution in blood? Imagine a happier hypothetical world in which blood had not been retributed? All English verbs, in all sentences, fall into one of three moods: indicative mood, imperative mood, and subjunctive mood.

Indicative mood is used for statements of facts, opinions, or ideas (such as in **declarative clauses**): *The man ate his hat. He likes it when men eat their hats.*

The indicative mood is also used to ask questions (such as in **interrogative clauses**): *Did the man eat his hat? The man ate his hat? Why did the man eat his hat?!* Word order can be reversed to reveal the speaker's intent to ask a question (*wh*-words and the question mark help too).

Imperative mood is used to demand something, prohibit something, or request something. The subject of the sentence usually drops out when we're giving a command in the imperative mood. These commands usually occur in the second person (singular or plural) form: *Eat your hat. Y'all eat your hats, now.* If the subject is invisible (a corollary in physics might be *virtual particles*, which are similarly there-but-not-there), then we mark that invisible subject with an *x* when we are diagramming: *(x) Eat your hat.*

Sometimes, however, the imperative is written in the first-person plural, as in *Let's go* (*Let us go*).

Subjunctive mood is used to make requests or suggestions, express wishes and desires, convey doubt, and describe hypothetical situations. These verb forms used to be much more common, but over the years, we've trimmed away many conjugations in English. Now we're left with only a few verb-form changes to signal to the reader that we're requesting something, hoping something, doubting something, or describing a hypothetical.

The **present subjunctive** is formed by using the base form of a verb—the infinitive without the *to*-particle. Since the base form of the verb is also used elsewhere, readers won't be able to tell you're using the present subjunctive tense unless your subject is singular. You'll see the present subjunctive mostly in some complement clauses beginning with *that* and in some adverbial clauses. Examples:

*She requested that he **arrive** alone.* (Normally *he arrives.*)

*Whether you **be** friend or foe, you are welcome in my house.* (Normally *you **are**.*)

The **past subjunctive** is used to express unreal or hypothetical meaning, and its form is only recognizable when the subject is singular. The past subjunctive is formed using the simple past tense of the verb (*if only I **had** Angelina Jolie's lips—but I don't*; compare the indicative mood version *I have Angelina Jolie's lips*), except in the case of *be*, which always takes the form **were** even if the subject is singular.

*If I **were** a duck, I'd be a wood duck because they're beautiful.* (If the situation were more probable, you'd use *was.*)

*I wish we **were** done with this chapter.* (But we're not.)

*If he **were** clever, he'd hire me immediately.* (A great, grammatically subtle way to insult someone!)

English has some great phrases left over from our more subjunctive days. There's "So be it," "Come what may," "Perish the thought," "God save the Queen," and "Suffice it to say," but my favorite used to be, "The devil take you!" It's not my favorite anymore.

Finally, **exclamatory clauses** can be in any of these moods. Exclamatory clauses end in exclamation marks. Use them sparingly.

Indicative mood: *The man ate his hat!*

Imperative mood: *Eat your hat!*

Subjunctive mood: *The devil eat your hat!*

Independent Clauses

An **independent clause** is a group of words that can stand alone as a complete sentence (that is, the core valency pattern contains at least one subject + one verb, and the clause doesn't start with a subordinating conjunction, complementizer, or relativizer). We just covered the types of independent clauses in these various moods: declarative clauses, interrogative clauses, imperative clauses, present subjunctive clauses, past subjunctive clauses, and exclamatory clauses. Now we'll talk about how to add clauses together in the same sentence.

Independent clauses can be grouped together (coordinated), and they can also contain embedded dependent clauses. Besides subjects and verbs, independent clauses can also have the other syntactic roles—direct objects, indirect objects, adverbials, and so on—but we'll keep it simple here. (Whenever you see S + V, know that you can substitute it with any of the more complicated valency patterns from Table 3.1.)

We'll look at four types of sentences now, organized by how many clauses they have in them: a simple independent clause, coordinated independent clauses, complex independent clauses, and the compound-complex sentence.

Simple independent clause (also called a **simple sentence**): one independent clause. *She sat* (S + V).

Coordinated independent clauses (also called a **compound sentence**): two or more independent clauses connected by coordinating conjunctions (remember FANBOYS?). *She sat, and she ate* (S + V, cc S + V). You can see an example of a compound sentence in Figure A.5 of the appendix.

When you have a compound sentence connected with one of the FANBOYS, make sure you put a **comma** (,) before the coordinating conjunction (S + V, cc S + V). This helps your reader realize there are *two* independent clauses at play, versus *one* independent clause where one subject fuses with two verbs—the compound predicate. (See *Chicago* 6.22 and 6.23 for more information.) While it used to be common to tack on a comma, a coordinating conjunction, and a second verb to an independent clause (S + V, cc V), doing so actually introduces the potential for miscommunication, especially if a second noun inside a multi-word subject matches the second verb. Clarify the clause by creating coordinated independent clauses (S + V, cc S + V) or a proper compound predicate (S + V cc V). Figure A.4 in the appendix shows a diagram of a compound predicate. (However, if the first independent clause is very short, then you don't need the comma.)

Also note that if you only have a comma between the two independent clauses (S + V, S + V), it can be considered a **comma splice**. This problem can easily be fixed by replacing the comma with a semi-colon (S + V; S + V). Or make it into a set of coordinated independent clauses (S + V, cc S + V) or a compound predicate (S + V cc V).

An independent clause with at least one dependent clause is called a **complex sentence**. *Because she was tired, she sat* (sc S + V, S + V). Or *She sat because she was tired* (S + V sc S + V). Notice that you will almost always put a comma if the subordinating clause comes first. In other types of complex sentences, you slot a dependent clause inside the independent clause, which is discussed in the next section.

Compound-complex sentence: at least two independent clauses and at least one dependent clause. *Because she was tired, she sat, and she slept* (sc S + V, S + V, cc S + V). This is just one iteration, but you get it. (For the curious, how to add electron-like punctuation to these sentence nuclei is covered in Table 4.1.)

Dependent Clauses: The Tricky Stuff

Dependent clauses are headache-bequeathers and my least favorite part of grammar. Because they're clauses with at least one subject and one verb—or because they're PIG monstrosities from Lesson II—dependent clauses get their own mini-valency line either below the independent clause or within a slot inside the independent clause. Nests with dependent clauses quickly become complicated.

While it's important to master dependent clauses, good writers realize it's often best to write simpler, more straightforward independent clauses, without too many nested dependent clauses. We'll look at **complement clauses**, **adverbial clauses**, **relative clauses**, **comparative clauses**, and a grab-bag of other clauses that don't fit in these categories.

Complement clauses are beastly, fascinating things that deserve their own book. Here's a simplified version: complement clauses complete, expound on, or rename the meaning of a previous word. You've already met the subject predicate (also known as the **subject complement**) and the object complement as syntactic roles, but now we're going to cram more than one word in those slots. In almost all the other slots, too.

Like a phrase, a complement clause starts with a head word, but this head word gets a lovely Latinate name: a **complementizer**. (Why do so many grammar words have to contain *c* or *p*, or both?) Complement clauses generally start with *that*, *if*, *whether*, or the *wh*-words.

You'll often see complement clauses hanging out with reporting verbs like *say*, *reply*, *tell*, *know*, *think*, *inform*, *admit*, *argue*, *answer*, and *advise*. Take the example *I said **that I needed sleep**. As a test to see if you have a complement clause, delete the complementizer. What's left should be a clause that makes sense on its own: *I needed sleep*. (This same test fails if you apply it to relative clauses.) Sometimes sentences drop the word *that* entirely, making diagramming an even more exhilarating challenge: *I said I needed sleep*. The complementizer has become a virtual particle—there but not there. In diagramming, we mark invisible words with an *x* or put them in parentheses.

Sometimes a complement clause (complementizer + S + V) slides right into the subject, extraposed subject, subject predicate, direct object, or object complement syntax slots. To see an example of a diagrammed complement clause—and to help you distinguish among all the varying uses of the Function word *that*, see Figure A.1 in the appendix.

In the following examples of complement clauses slotted into syntactic roles, I've set the clause in bold and underlined the word or phrase that the clause is complementing, expanding, or renaming. Since almost all of these clauses act like nouns in those syntactic roles, you'll often see the term **noun complement clause** or **noun clause**. What's easiest to me is to fill in the blank: "This complement clause is acting as [word class or syntactic role]." Examples of complement clauses:

As subject: **What happened last Sunday** is an <u>outrage</u>. **That this was my fault** became <u>obvious</u>. **How it will end** may be <u>beyond my control</u>.

As extraposed subject (with the dummy pronoun *it*): *It is an* <u>outrage</u> **what happened last Sunday**. *It became* <u>obvious</u> **that this was my fault**. *It may be* <u>beyond my control</u> **how it will end**. (Science writers overuse this construction.)

As subject predicate: *The bad* <u>news</u> *is* **that she will return**. *The* <u>solution</u> *isn't* **what I wanted**.

As direct object: *He doesn't* <u>know</u> **whether the cat will come back**. *I will* <u>do</u> **what I must**.

As object complement: Dyna called <u>me</u> **what some might consider rude**.

Sometimes a complement clause renames another noun like an Appositive Phrase does, so in diagramming, you'd put the appositive-like complement clause in parentheses and place it on a stem next to whatever noun it renames. You'll see these kinds of complement clauses with reporting nouns like *claim, response, remark, comment, idea, fact, argument, assumption,* and *possibility.* Yes, science writers definitely overuse this kind of complement clause, too.

> As appositive: *Such knowledge challenges my <u>assumption</u> **that muses are good**.* The <u>fact</u> **that I can't sleep** annoys me.

What if the complement clause doesn't fit in any syntactic role? If you can't slide the clause into any syntactic role along any **valency axis**—the horizontal line of syntax in a Subatomic Writing diagram, as shown in Figure A.2—then the clause must be some kind of modifier. As a modifier, the complement clause will go *under* a valency axis as the object of a preposition or as an adverb that modifies an adjective:

> As object of the preposition: *I was afraid <u>of</u> **what might happen to me**. Let's talk <u>about</u> **why she did this**.*

> As adverb modifying adjective: *He was not <u>aware</u> **that it was my fault**. I'm not exactly <u>sure</u> **when she'll be back**.*

What about a complement clause acting as an adverb but modifying a verb? This kind of complement leads us to a whole new world of dependent clauses: the adverbial clause.

Adverbial clauses modify verbs, usually the main verb in an independent clause. Adverbial clauses start with a head word too—remember subordinating conjunctions?—and they have a subject and a verb, which is why they're called clauses. You'd diagram an adverbial clause (sc S + V) with a slanted dotted line under the main verb, with the subordinating conjunction written along the dotted line. They can answer the same questions as adverbs do.

> *__Before you move in__, I'd <u>like</u> to get rid of some junk.*

> *__When you're ready__, <u>bring</u> the car around, **although you'll have to park up the street**.*

A **relative clause** modifies a noun or noun-like entity: *The cat, whom I love, is gone.* Whereas a complementizer signals a complement clause and a subordinating conjunction signals an adverbial clause, a **relativizer**—a *wh*-word or the pronoun *that*—must play a syntactic role itself in the relative clause. In the relative clause *whom I love*, the relativizer is *whom.* To find out which role it's filling (and to diagram it), you have to untwist the clause into the regular valency pattern order: *I love whom*, and a vertical dotted line will connect *cat* and *whom.* You can see examples of diagrammed relative clauses in Figures A.1 and A.6. If the relativizer is the pronoun *that*, remember that it might become invisible like the complementizer *that.*

A relativizer's antecedent (the noun or noun-like entity it points back to) is often the subject of the independent clause, like *student* is in this sentence: *The student, who likes grammar, gets up early to make sentence diagrams.* Note that this example is a **nonrestrictive** relative clause talking about one specific student, and this one student happens to like grammar so much they get up early to study it.

Without the commas, the sentence now has a **restrictive** relative clause: *The student who likes grammar gets up early to make sentence diagrams.* This now means that any student who truly likes grammar should or will get up early. Restrictive and nonrestrictive words, phrases, or clauses are tough, so if you don't quite get them yet, review Appositive Phrases in Lesson II or spend a few minutes on Google.

Comparative clauses act as complements in an Adjective Phrase or an Adverb Phrase, and they have a gradable word as the head word. Examples:

The comparative clause can be part of an Adjective Phrase: *I am not [as clever **as I pretend to be**].*

The comparative clause can be part of an Adverb Phrase: *He keeps leaving work [earlier **than he should**].*

To wrap up, **peripheral clauses** refer to the group of seven odd-ball clauses that don't fit well into any other classification.

- **Reporting clauses** are on the boundary between independent and dependent clauses. They're classified as dependent clauses, however.

  "I'm going to eat you," **the monster said.**
  She screamed at him, *"Don't touch my coffee!"*

- **Question tags** are loosely attached to another clause, usually at the end.

  We're done here, **aren't we?**

- **Supplement clauses** are loosely connected to the main clause.

  Considered the last true princess of the Korean Empire, *Princess Deokhye was born on May 25, 1912.*
  He stared across the room, **eyes glazed.**

- **Verbless clauses** can be treated as adverbial clauses with a missing subject and the verb *be*.

  When [you are] in trouble, *think ahead.*
  Every week, **if [it is] possible,** *sit down with your schedule.*
  Whether [we are] early birds or night owls, *we should get enough sleep.*

- *Ing*-**clauses**, *ed*-**participle clauses**, and **Infinitive clauses** (also considered complement clauses) are nothing more than

Participial, Gerund, and Infinitive Phrases by another name! Since we vanquished the PIG in Lesson II, all we have left to do is to keep practicing them. Congratulations!

English grammar is tough, no question, particularly if English is not your first language. The word *grammar* itself comes from the Greek *grammatikè téchnē*, which means "art of letters," from γράμμα (*grámma*), "letter," which itself comes from γράφειν (*gráphein*), "to draw, to write."

CONCISION AT THE SYNTAX LEVEL

One final thought on generating light and understanding for your reader at the syntactic level. Decay is important throughout any stage of Subatomic Writing: we want our writing to be concise, with phrases and clauses arranged to give the clearest picture possible to the reader. We won't always succeed, but we can sure try.

As the poet Robert Southey said, "It is with words as with sunbeams. The more they are condensed, the deeper they burn." While we don't want to give the reader gamma-ray burns, we do want to provide them with concise, clear constructions that provide optimal light, warmth, and energy.

EXERCISE 1

Diagram any of these questions in your preferred sentence diagram system. Hint: Rephrase the question as a statement.

Where is my cat?
When will my cat return?
How will he get back?
For what reason would she appear?
Why is she here?
Which books are best?
What time is it?

Who does not call Dyna a genius?

With whom are you talking?

Further, what sort of movement is primary?

    —Aristotle, "The First Cause," *Metaphysics*

What hinders the fixed stars from falling upon one another?

    —Isaac Newton, *Optics*, Book III, Part I, Query 28

EXERCISE 2

Diagram these **intransitive** sentences (S + V).

Stars burn.

Cats will be saved.

Quarks may have been colliding.

The wind had been blowing on that October night.

Grinning, Maxwell's mother's one blue vulgar demon recently
sat heavily on her to wake her.

Moon dust clings like dryer socks.

    —Mary Roach, *Packing for Mars*

His lungs were charged with his silence.

    —Frank Herbert, *Dune*

Hint: The verb *were charged* is a passive-voice verb. Rewritten
in active voice, the sentence would read, *Silence charged his
lungs.* The passive voice gives the sentence an S + V valency pattern and demotes the subject to the object of a preposition
(OP). The active voice, in contrast, gives the sentence an
S + V + DO valency pattern, rethroning the subject.

Never be limited by other people's limited imaginations.

    —Mae Jemison

By the pricking of my thumbs, something wicked this way comes.

    —Shakespeare, *Macbeth*

Clutching our crystals and nervously consulting our horoscopes, our critical faculties in decline, unable to distinguish between what feels good and what's true, we slide, almost without noticing, back into superstition and darkness.

    —Carl Sagan, *The Demon-Haunted World: Science as a Candle in the Dark*

EXERCISE 3

Diagram the less common **copular** pattern (S + V + A).

The demon is here.
The conversation lasted into the night.

EXERCISE 4

Diagram the more common **copular** pattern (S + V + SP). Remember that the subject predicate can be either an adjective (linking the subject with a characteristic) or a noun (renaming the subject). While this valency pattern is potentially powerful, we overuse it in science writing and writing in general.

I am she.
My editor is amazing.

The subject of our inquiry is substance.

    —Aristotle, "The First Cause," *Metaphysics*

The investigation of the meaning of words is the beginning of education.

    —Antisthenes, c. 445–c. 365 BCE

Comparative:

The universe is wider than our views of it.

—Henry David Thoreau, *Walden*

Hint: The word *than* is a relative adverb modifying the adjective *wider* and introducing an elliptical adverb clause—a clause where pieces are missing but our brains fill them in. The full, undeleted sentence would be: *The universe is wider than our views of it (are) (wide).*

Infinitives:

To be rational is to look the universe in the face and not flinch.

—Unknown

To know

That which before us lies in daily life

Is the prime Wisdom.

—John Milton, *Paradise Lost*

Education is the most powerful weapon, which you can use to change the world.

—Nelson Mandela

Brevity is the soul of wit.

—Shakespeare, *Hamlet*

Defining myself, as opposed to being defined by others, is one of the most difficult challenges I face.

—Carol Moseley-Braun

Hint: *Defining* is a gerund, and *myself* is the direct object of the gerund. This Gerund Phrase (GP) is the subject in this sentence. There is also an invisible *that* after *challenges*. *That* is a pronoun and part of a relative clause.

The limits of my language mean the limits of my world.
  —Ludwig Josef Johann Wittgenstein

It is probably no exaggeration to say that all of theoretical physics proceeds by analogy.
  —Jeremy Bernstein, *Elementary Particles and Their Currents*
Hint: This sentence has what's called an extraposed subject, where the real subject is kicked down the road and replaced with the dummy pronoun *it*. The sentence really reads: *To say that all of theoretical physics proceeds by analogy is probably no exaggeration.*

EXERCISE 5

Diagram these **monotransitive** sentences (S + V + DO).

She dropped my book.

Compound direct object:
She dropped my books and my glass tumbler on the floor.

Whole clause as direct object:
Much learning shows how little mortals know.
  —Stephen Clark, *A Practical Grammar*

Transitive sentences with Appositive Phrases:
Electrons know two verbs: seek and avoid.
  —Felice Frankel and George M. Whitesides, *On the Surface of Things*

Quantum mechanics taught that a particle was not a particle but a smudge, a traveling cloud of possibilities.

> —James Gleick, *Genius: The Life and Science of Richard Feynman*

Now all things that change have matter, but different matter.

> —Aristotle, "The First Cause," *Metaphysics*

Clear your mind of can't.

> —Samuel Johnson, 1709–1784

Nature possesses an order that one may aspire to comprehend.

> —Chen-Ning Yang, Nobel Lecture, 1957

EXERCISE 6

Diagram these **ditransitive** sentences (S + V + IO + DO). In my opinion, the ditransitive pattern is not much more than a variation of S + V + DO. The indirect object usually happens after verbs like *tell* and *give*.

I won't give Max a phone yet.
Don't tell your reader what to feel.

EXERCISE 7

Diagram these **complex transitive** sentences (S + V + DO + OC). The secret to identifying the object complement (sometimes called the object predicate) is to realize that without the OC, the sentence doesn't make sense. It feels incomplete without the needed OC (or the obligatory adverbial, A, below).

Earth's atmosphere keeps us safe.

She makes me so mad.

Make not your thoughts your prisons.
    —Shakespeare, *The Taming of the Shrew*

I laid my heart open to the benign indifference of the universe.
    —Albert Camus, *The Stranger*

A less common kind of **complex transitive** construction is
S + V + DO + A:
I should have sent her away.
I placed the notebook and pencil on the coffee table.

Don't put a ceiling on yourself.
    —Oprah Winfrey

EXERCISE 8

Diagram these **compound** sentences (variations of S + V, cc
S + V).

The new burn on the carpet is bad, but I managed to cover it up.

I came, [and] I saw, [and] I conquered.
    —Julius Caesar

Just get it down on paper, and then we'll see what to do with it.
    —Maxwell Evarts Perkins, 1884–1947

EXERCISE 9

Diagram these **complex** sentences (Some combination of sc
S + V, S + V).

Though this be madness, yet there is method in't.
    —Shakespeare, *Hamlet*

Weighest thy words before thou givest them breath.
    —Shakespeare, *Othello*

Though thou speakest truth, methink thou speak'st not well.
    —Shakespeare, *Coriolanus*

EXERCISE 10

Diagram these **compound-complex** sentences (some form of sc S + V, S + V, cc S + V).

I sighed, and while I understood, I didn't like it.
"How silly," I thought, and I got back in bed, even though part of me still feared sleep.

When you get right down to it, everybody's having a perfectly lousy time of it, and I mean everyone.
    —Kurt Vonnegut, *Sirens of Titan*

EXERCISE 11

Using Table 3.1, identify what valency patterns these sentences use in any type of clause they might have (independent, dependent, relative, etc.).

The vibrations set in motion by the words that we utter reach through all space, and the tremor is felt through all time.
    —Maria Mitchel, in Helen Wright, *Sweeper in the Sky*

We know light best in its diluted form: a gentle rain of photons falling from the sun that illuminates and warms. More concentrated, light is a furnace and a terror.

—Felice Frankel and George M. Whitesides, *On the Surface of Things*

modifiers

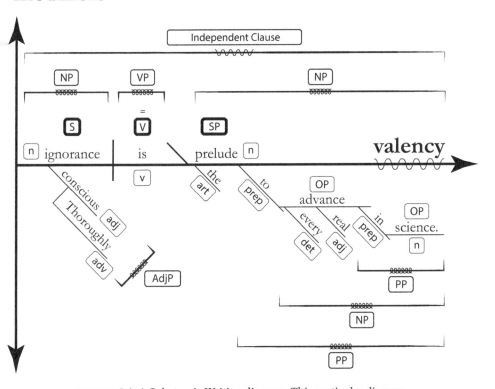

FIGURE 3.4 A Subatomic Writing diagram. This particular diagram incorporates word class from Table 1.1 (noun, adjective, etc.), one word or multi-word phrases from Table 2.2 (Noun Phrase, Verb Phrase, Prepositional Phrase, etc.), and a valency pattern from Table 3.1 (Subject + Verb + Subject Predicate). You can see how the recursive nature of nests within nests happens, even in this fairly straightforward thought from James Clerk Maxwell: *Thoroughly conscious ignorance is the prelude to every real advance in science.*
Source: Jamie Zvirzdin.

Though in the course of ages catastrophes have occurred and may yet occur in the heavens, though ancient systems may be dissolved and new systems evolved out of their ruins, the molecules out of which these systems are built—the foundation-stones of the material universe—remain unbroken and unworn.

—James Clerk Maxwell, *The Scientific Papers of James Clerk Maxwell*, Volume 2

The "mare" in "nightmare" originally referred to a demonic woman who suffocated sleepers by lying on their chests (she was called "Old Hag" in Newfoundland).

—Oliver Sacks, *Hallucinations*

EXERCISE 12

Once you feel good about your Reed-Kellogg skills, try making a Subatomic Writing diagram, labeling all that you have learned so far. Check out the "helium-3" sentence from James Clerk Maxwell in Figure 3.4. Additional diagrams and guidelines for creating Subatomic Writing diagrams can be found in the appendix.

MIND'S BREATH

Sentence Level

> Punctuation is the art of marking in writing the several pauses that take place in speech.
>
> —John Dalton, *Elements of English Grammar*

CONGRATULATIONS, you have survived Lesson III, with all those terrible clauses. I've made Lesson IV short and easy, at least in comparison. Many students who have issues with **punctuation, emphasis**, and **spacing**—the literary leptons introduced in Table 0.2 —actually struggle with word classes and syntax patterns, so good news—you've already done that work. In this lesson, we'll briefly introduce the Z^0 boson and cover the basics of English typographical conventions, starting with the easiest and ending with the hardest. Our goal is to take the nuclei-like clauses from Lesson III and form complete sentences—stable literary atoms. First, however, John Dalton makes an unexpected reappearance.

JOHN DALTON, THE GRAMMARIAN

You can imagine my astonishment when, digging around the internet, I learned that John Dalton—father of atomic theory (Figure 4.1), builder of wooden spheres linked into clusters, butt of demon jokes—also wrote a grammar book (Figure 4.2).

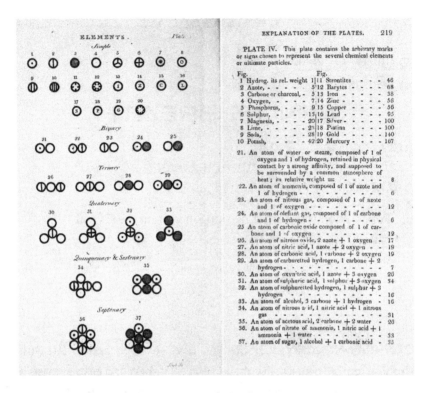

FIGURE 4.1 The beginnings of visualizing atoms as spheres.
This was written after Dalton published a book on
the particles of language in 1801. Coincidence? Maybe.
Source: John Dalton, *A New System of Chemical Philosophy*,
Part I. Manchester: S. Russel, 1808, Plate 4 and p. 219.

Dalton clearly knew the rules of grammar for his time, knew them
so well he sought to teach them to others. Besides loving chemistry,
lawn bowling, recording the weather, and drinking down at the Dog
and Partridge pub just outside town, Dalton was a prolific writer, and
he also acted as secretary for the Literary and Philosophical Society of
Manchester. He edited many papers and obsessed over errors, creating thorough "Errata" sections. In his diaries, he recorded how he'd
read Isaac Newton's *Principia* in the original Latin. This grammarian-
scientist expertly navigated the particles of his literary universe, which
improved his ability to navigate the particles of the physical universe,

ON PUNCTUATION.

Punctuation is the art of marking in writing the several pauses that take place in speech.

The points or stops used for this purpose, are:—
The COMMA (,) denoting a momentary pause;—
The SEMICOLON (;) denoting a pause about twice as long as a comma;—
The COLON (:) denoting a pause twice as long as a semicolon;—and
The PERIOD (.) denoting a pause twice as long as a colon.

The COMMA.

1. When several words of the same sort or part of speech follow one another, the comma separates them, except the two last, which have usually and, or, &c. between them, without the comma: as, "A character compounded of ignorance, pride, avarice and impiety;" "She is a beautiful, sensible, polite and amiable woman;" "He respects, esteems and admires her;" "They entreated us earnestly, tenderly and affectionately;" "By, with or through what influence he has acted, does not appear."

2. After the nominative case absolute, and on each side of short, intermediate and relative sentences, a comma is placed; as, "The time of youth being precious, it ought not to pass without improvement;"—"No person, who has any regard for his character, would take part in such a business."

3. Nouns in apposition usually have a comma before and after them; as, "Thomson, the poet, has finely described it."

4. A comma is placed after the name or title of a person spoken to; as, "John, come hither;" "My son, attend to my instruction."

5. A comma is put before the conjunction AND, but not before the preposition AND:—as, "You may stay here, AND I will go yonder;" "Two AND two make four."

6. Lastly, a comma may be inserted before most conjunctions when they are used as such, that is, as connecting or affecting sentences.

The SEMICOLON.

A long and compound sentence is generally broken into two or more parts by semicolons; they are placed in situations where the connection is not very close, and where a greater pause than a comma is admissible; a due attention to the sense and construction will be a sufficient guide in this respect.

The COLON.

When a sentence of considerable length is complete in itself, as far as respects the construction, but a small addition is made to it; then the part added may be separated by a colon: as, "We have, at his request, collected all the information we could procure upon the subject, and have transmitted it to him : it remains with him to determine."

FIGURE 4.2 A spread from Dalton's *Elements of English Grammar*. As you can see, punctuation usage continues to shift around through the years. If you google "Library UPenn Dalton Elements of English Grammar," you can read the original book. Dalton does not mention demons—I checked. It's also a little outdated according to modern-day American English usage.

Source: John Dalton, *Elements of English Grammar* [. . .]. London: R. & W. Dean, 1801, pp. 118–19.

especially when it came to expressing his ideas clearly and collaborating with others. The scientists we remember the most, like Dalton, are often the ones who knew how to write well.

Z^0 BOSON

Lessons IV and V use the **weak force** as our representative exchange particles, which come in three types, the Z^0 boson and two W bosons, W^+ and W^-. (The Z of the Z^0 boson stands for "zero-charge weak force," in contrast to the positive and negative charges of the W bosons. The

W ingeniously stands for "weak.") The weak force carried by these three exchange particles is not as powerful as the strong force and the electromagnetic force (gluons and photons). But the weak force does govern how quarks and leptons move together or apart, like the context-driven interactions between grammar, typographical conventions, and rhythm (we'll get to rhythm in Lesson V).

While a case could be made for using the W bosons for this lesson, I chose the Z^0 boson. Its neutral force interacts between quarks and leptons, and since this boson has mass itself, it can decay into photons. In the literary universe, typographical conventions similarly interact between and among words, turning into additional illumination for the reader. Z^0 can also decay into an electron and an antimuon—or the reverse, a positron (an antimatter electron) and a muon. The ATLAS collaboration at CERN says that if Z^0 decays into a tau lepton, it could also produce an electron, muon, and other particles. There is a mishmash of options, as you can see, and we're still figuring out what goes where, who affects what—and why. The same can be said of the extra layer of language we use in writing, since publishers, editors, and authors still debate about best practices in punctuation, emphasis, and spacing.

Or, as you see in Figure 4.3, the Z^0 boson is also the go-between that causes a change in momentum between a neutron and an electron neutrino. If neutrons are like phrases and electron neutrinos are like spaces between words, the typographical conventions that govern their interactions (in our case, *Chicago* style) describe how and when to use tiny spaces to change the momentum of your words and bring further substance and energy to your message. Suffice it to say, there are a lot of options when it comes to choosing which typographical conventions best fit your writing needs. I've learned many different style guides over the years, but I find that *Chicago* provides the most clarity of thought. But as my friend Mary Price—a brilliant violinist and mathematician—likes to say, "There's no sense arguing over a matter

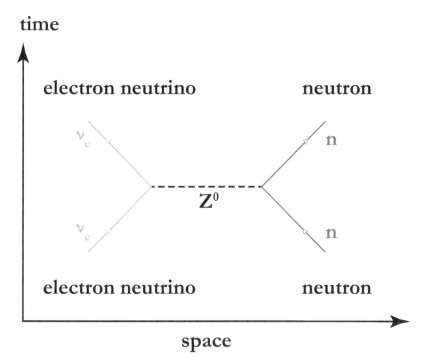

FIGURE 4.3 A Feynman diagram of the Z^0 boson in action. Because this boson doesn't exchange charge, it won't change the nature of the electron neutrino and the neutron; however, it will change their momentum.
Source: Jamie Zvirzdin.

of taste." If your publication of choice uses a different style, using *Chicago* as your base will still help you differentiate all the little lepton-like bits they ask you to include in your final manuscript.

NEUTRINOS: SPACING

We use different weights of spacing among the levels of language to communicate extra meaning to the reader: between words at the phrase level (the literary **electron neutrino**), between sentences at the clause level (the **muon neutrino**), and between paragraphs at the paragraph level (the **tau neutrino**). Neutrinos are small, ghost-like particles, but they still contribute a great deal to our physical universe. The same can be said for literary neutrinos.

On **word-level spacing**: If you struggle to know whether a compound word has a space between it ("compound"), needs a hyphen ("hyphenated"), or shouldn't have any space at all ("closed up"), you can understand the basic rules by reading *Chicago*, Sections 7.81–7.89. It might be easiest to search for the word online at www.m-w.com (*Merriam-Webster*, the dictionary *Chicago* uses) to see the treatment of the word.

On **sentence-level spacing**: Most publishers use **one space** between sentences now. I retaught myself to type one space instead of two, although it's always smart to use the find-and-replace feature in your writing software to check for extra spaces—editors can spot them a mile away, and they can be distracting to the reader.

Some institutions, like the US State Department, have internal writing styles that differ even in this: when he's writing a diplomatic cable, my husband is required to use two spaces between sentences— for long, historically sordid reasons that I think are funny.

When your sentences wax too long and too complicated, readers run out of mental air and need to take a breath. They need this space to keep up with you. The average English sentence is fifteen to twenty words long, but we aim for a healthy mix of sentence lengths in any one paragraph.

On **paragraph-level spacing**: The **pilcrow** is the invisible symbol, ¶, that marks the start of a new paragraph, along with any indentation you might add. Beginning a new paragraph with this more massive bit of space is an underappreciated subatomic signal to the reader that you're moving to the next point. It allows the reader a longer mental breath, which we need sometimes to survive dense information in writing. You can greatly increase the readability of your writing by managing how long your paragraphs are (the average paragraph length is three to five sentences). This variety is called **pacing**, and it's one of the skills we'll cover in Lesson VI.

MUONS AND TAUS: INDIVIDUAL
AND WHOLE-WORD EMPHASIS

Any kind of emphasis helps draw your reader's attention to the most important parts of your message (too much, however, can be overwhelming). I won't be spending much time on **muons** and **taus**, as fascinating as they are in the physical universe: these particles in Subatomic Writing (**individual** and **whole-word emphasis**) vary from publication to publication, although they certainly make a difference in writing too.

If you struggle with capitalization, italics, boldface, and quotation marks (another form of emphasis that we cover in the punctuation section), I recommend reading *Chicago*, Sections 7.48–7.80. For conventions for using superscripts and subscripts, see Sections 12.36–12.38. *Chicago* uses uppercase letters (**capitals**) sparingly, usually to mark proper nouns, in what is known as a "down style" (*President Lincoln* was a great *president*).

However, *Chicago* also prefers headline-style capitalization of titles, which is called "up style" (Section 8.1). A good rule of thumb for capitalizing titles: Capitalize the first letter of the first word, all Lexical words, and all pronouns, primary auxiliaries, and modal auxiliaries. (See how word class comes back to haunt us?) Don't capitalize the other word classes like articles (*a, an, the*) and prepositions (*in, around, of*) and coordinating conjunctions (FANBOYS). The titles of major works like books, journals, and plays are italicized while subsections of larger works like essays, articles, poems, and book chapters are set in quotation marks. Underlining, as a form of emphasis, is rarely used these days except in textbooks when extra emphasis is needed to distinguish meaning.

ELECTRONS: PUNCTUATION

Let's take the deuterium nucleus again, which, like sun gods, we created in Lesson III: the nucleus of deuterium is one proton fused to one neutron. It is not an atom, however, until it also has one **electron**—only then is its charge balanced and stable. We likewise need punctuation to complete our clauses and stabilize a full sentence.

If you remember electron orbital shells from chemistry class, you might remember that these "probability clouds" represent where an electron might be found around a nucleus 90 percent of the time. The number of electrons an atom has determines how many orbital shells it possesses. Deuterium has the same **electronic configuration** as hydrogen: $1s^1$. This means there is one proton and one electron in the first orbital shell, *s*. Like a period completing a sentence, an electron completes the atom.

The more complex a sentence, the more punctuation you'll need to fill those orbital shells and balance the additional phrases and clauses in your nuclei. Your electron-like punctuation should seek out places in your sentences to **pause**, **link**, and **tighten**, bringing balance to your sentences and further light and substance to your reader. We will cover these methods later in this lesson, but first, it helps to remember where punctuation came from and why we use it in the first place.

PUNCTUATION, A BRIEF HISTORY

I fully sympathize with people who hate punctuation because it is capricious—it seems you never know precisely where you might find all those little dots, just as electrons have probabilistic orbital shells around the nucleus of an atom. Punctuation is practically a separate language, adding even more dots and lines to a language that itself is a tangled mess of languages. Whether you feel passionately about Oxford commas and em and en dashes—or feel nothing at all for them, or fully resent this additional layer of language—remember to keep the purpose of punctuation at the forefront: clarity for the reader.

Clarity is the reason punctuation and spacing were created in the first place—to make the flow of words easier for the reader to understand. Half the battle for readers was trying to figure out what the words were. The earliest usage of punctuation we have found is written on a basalt block called the Mesha Stele, from 840 BCE. The author put a dot between words and a horizontal stroke between "sense sections," or what we might call paragraphs.

Then, around 200 BCE, Aristophanes of Byzantium, a Greek librarian in Alexandria, used a dot, placed either at the top, bottom, or middle of the line after a word, to indicate different lengths of speech pauses. The top dot stood for **the longest pause** and became our period. The bottom dot was a **medium pause**, what became our colon and semicolon. The middle dot was a **short pause**, what became the comma.

Romans didn't adopt Aristophanes's dots, so the shouty caps Dyna mentioned continued until the seventh century CE, when Isidore of Seville resurrected the system of dots. In *Etymologiae*, he wrote about all kinds of great topics, including geometry, animals, music, and grammar. Then Dante and Chaucer got on board, and the punctuation system spread and evolved from there. The book *Shady Characters: The Secret Life of Punctuation, Symbols & Other Typographical Marks*, by Keith Houston, lays out the remarkable history of how lepton-like conventions arose in our language, if you're interested. Houston's book had the odd effect of generating a deep sense of gratitude in me toward . . . punctuation. A writer who can do that has skill.

The three main ways punctuation will help you amplify your message is by pausing, linking letters, words, phrases, and clauses, and tightening your message with subtle concision tricks.

PAUSING

As Dalton's quote says at the beginning of this lesson, different punctuation shows the pauses in speech. Along with emphasis and spacing, these pauses interrupt strings of letters so our minds can breathe. From Lesson I, we learned that a reader still "hears" your writing, even if they're reading silently, so punctuation still has this speech-like effect on the reader. When done well, the reader can hear you pause, and they can even tell if you intend a long, medium, or short pause. (These categories are also listed in Table 0.2.)

Long Pauses

. = **Period.** The speaker's voice drops, finishing a complete thought. An easy, subatomic action that improves the quality of a finished product is to make sure you have a period at the end of each sentence. Nobody's perfect, but double-check your writing for this issue before submitting it.

? = **Question mark.** The speaker's voice goes up at the end (unless you're British), seeking information, clarification, or any kind of response.

! = **Exclamation mark.** The speaker's voice gets louder or more forceful, emphasizing something.

Medium Pauses

; = **Semicolon.** The speaker's voice pauses. This is a slightly shorter pause than a period because the two independent clauses the semicolon connects are logically linked to each other. Less common usage: to link a series of long items in a list.

: = **Colon.** The speaker's voice pauses before introducing a list, a definition, a summary, or an important conclusion.

— = **Em dash.** The speaker's voice pauses, for effect or to add a related but more parenthetical side thought. The full dash—like this—looks and flows much better than two hyphens (--), the en dash (–), or a single hyphen (-). In *Chicago*, there are no spaces on either side of the em dash—like this. Take a moment to look up and memorize the shortcut command for the em dash on your keyboard (I press Option + Shift + Hyphen). It is so much easier to use if you treat the em dash like another letter on your keyboard. You will save yourself loads of time. Same with the en dash (–), whose shortcut, at least on my computer, is Option + Hyphen.

. . . = **Ellipses.** The speaker's voice trails off, with the expectation that the audience can figure out from context what might have been said next. In *Chicago*, there's a space between each of the three (sometimes four) periods. Four periods, when citing a source, indicate that you're skipping across sentences.

Short Pauses

, = **Comma**. The speaker's voice briefly pauses, for one of many reasons. The comma is the hardest mark to master. If you can stomach it, I recommend reading *Chicago*'s section on comma usage, Sections 6.16–6.55. When we speak, we use different kinds of short pauses:

- Before a coordinating conjunction that connects two independent clauses: *S V, and S V*. If you write *S V, and V*, then you have a problem: the compound predicate problem. Don't put a single comma between a subject and any of the verbs it belongs to. To fix the compound predicate problem, either repeat the second subject in a compound sentence construction (*S V, and S V*) or delete the comma (*S V and V*) for a proper compound predicate construction.
- After an introductory phrase or dependent clause: *When it rains, it pours.*
- To separate items in a series, including the "Oxford comma"— the comma that comes after the second-to-last item and before the FANBOYS coordinating conjunction: *Quarks, leptons, and bosons are the three kinds of particles in our current Standard Model.*
- To set off a parenthetical or nonrestrictive phrase: *Dalton, who loved chemistry, also wrote a grammar book.*
- Between the day and year of a date: *October 31, 2021.*
- Before and after a named state or country: *The odd weather in Houston, Texas, is concerning.*
- To set off quotations within a sentence: *She said, "You must make time for the things you want in life."*
- To indicate direct address: *Hey Dyna, I don't deserve this.*

- To separate noncumulative adjectives: *The unusual, untidy, Greek demon arrived.*
- To indicate omissions of verbs in parallel clauses: *I like hot cocoa; Max, orange juice; and Dyna, Scotch.*

A critical test of your punctuation skills is to read your sentence out loud: **Do you run out of breath before you reach the end of the sentence?** If so, you either need to break the sentence into two sentences or do a better job organizing your phrases and clauses with punctuation. Remember, readers read-hear your words in their heads; don't deprive them of mental oxygen.

Conversely, if you read your sentence out loud and find you're pausing in strange places, like between a subject and verb (where commas shouldn't go), then take out the punctuation. You don't want the reader to run out of mental breath when they read your words, but neither do you want to make them hiccup through a minefield of overly punctuated words.

LINKING AND TIGHTENING

Another function of punctuation, beyond pausing, is to establish additional relationships between words and increase the pace by shortening words. Both of these functions aid reader comprehension.

Common Punctuation to Link and Tighten

'= **Apostrophe.** We learned about using the apostrophe to show possession (*one demon's victim*; *two demons' victim*) back in Lesson I, but you can also use apostrophes to link and tighten two words into one **contraction**: *I am → I'm. They had → they'd. We have → we've.* Contractions are less formal, but they are often perfectly acceptable in science writing. They can speed up the pace of a sentence.

- = **Hyphen.** The hyphen causes endless headaches. I still have to look up *double-check* (the verb) in the dictionary because I never remember if the verb has a hyphen or not (it does). Remember to link compound attributive adjectives with a hyphen: *the **three-eyed** demon*. You know an Adjective Phrase needs a hyphen if the words don't make sense alone: *The three demon. The eyed demon*.

. . . = **Ellipses**. Ellipses can also show that you omitted a part of a quotation unrelated to your topic. There are specific rules in how and when you can cut out parts of a quote to tighten up the point you're making. *Chicago*, Sections 13.50–13.58 will give you the nitty-gritty details—including the use of four periods (13.53).

. = **Period**. This mark is used to shorten—abbreviate—terms your reader will know. In *Chicago*, we use ***et al.***, which is short for Latin *et alia*, "and others." We also use ***e.g.,*** and ***i.e.,*** (notice those periods and commas) to stand for *exempli gratia*, "for example," and *id est*, "that is." Whenever possible, just write "for example" and "that is." The more abbreviations and acronyms you have in a text, the longer it will take the reader to unpack the abbreviation, interpret the dots, and move on. Too many electrons in one place is a negative thing. *Chicago* has an entire section on abbreviations, Section 10.

More Linking Relationships

These additional types of marks signal to the reader that words within or immediately around these punctuation marks are either closely related or loosely related to other words nearby.

( ) = **Parentheses**. Parentheses set off a loosely related phrase.

[ ] = **Brackets**. Brackets show inserted words not present in the original source.

– = **En dash**. The en dash has three functions: first, to replace the word *through* in a range of inclusive numbers or months: *pages 30–40, years 1903–1976, the budget for January–April 2003*. Second, the en dash replaces the hyphen in compound terms when one element of the compound is itself a hyphenated or an *open* (nonhyphenated) two- or three-word element: *the post–World War I recovery, a Los Angeles–based* *company, La Niña–related damage, the North Dakota–South Dakota squabble.* Finally, the en dash is used to report scores or tallies: *The court split 5–4.* **NOTE: In a bibliographic entry or source citation, en dashes are often used instead of hyphens for page number ranges. Editors take notice because they're often the ones who have to fix them.**

" " = **Double quotation marks**. Use these to quote material from another writer. Don't use them as "scare quotes," unless you're meaning to be sarcastic. They are also used to represent text as speech, as well as to signal titles of poems, short stories, articles, names of specific webpages, etc. Section 14 of *Chicago* covers all the nitty-gritty glories of creating notes and bibliographic entries. **NOTE: In American English, periods and commas always go inside quotation marks. Question marks, exclamation marks, semicolons, colons, and dashes go outside quotation marks unless they are part of the quotation.**

' ' = **Single quotation marks**. If you're quoting someone who is in turn quoting someone else, use the singles inside the doubles: *She said, "I heard someone saying, 'Who is there?'"*

NOTE: If you copy-and-paste material, you can accidentally introduce *straight quotes* into your text instead of the preferred *curly quotes*. Editors prefer curly quotes for publishing. You can quickly make sure your quotation marks and your apostrophes are curly by using the find-and-replace feature in your word processor (just use the same mark in both fields, and all the straight quotes will turn into curly quotes).

, [words], = **Nonrestrictive phrase signal**. Putting two commas around a word, phrase, or clause means that what's inside the commas is **nonrestrictive**—extra information that if taken out, wouldn't destroy the meaning of the head noun it connects to. This is often how Appositive Phrases are punctuated and how you can recognize that the appositive is renaming or providing additional (nonessential) information about the preceding noun.

/ = **Solidus**. This mark indicates multiple possibilities, as well as line breaks in quotations of poetry.

This is not an exhaustive list of all possible punctuation or types of typographical marks, but it can help you discern the purpose of mainstream punctuation.

Becoming a master of punctuation means knowing approximately—in which orbital around the sentence—a dot or dash goes. Table 4.1 shows how to combine syntax patterns, punctuation, and spacing to form sentences from simple to complex: fragments, simple sentences, compound sentences, complex sentences, and compound-complex sentences. It is a veritable periodic table of sentence patterns.

I will leave it to chemistry lovers to see if they can somehow match these sentence patterns and their variations to the 118 chemical elements of the period table, or according to an atom's azimuthal quantum number, from 0 to 6 (historically called by the orbital shell letters *s*, *p*, *d*, *f*, *g*, *h*, and *i*). In *Make It Stick: The Science of Successful Learning*, Brown, Roediger, and McDaniel state, "Elaboration is the process of finding additional layers of meaning in new material. . . . A powerful form of elaboration is to discover a metaphor or visual image for the new material." Whether we like it or not, the sillier or more exaggerated a metaphor, the more likely we are to remember it (the "mind palace" method of memorization, by the way, also supposedly came from ancient Greece).

For additional practice and examples of sentence patterns, I recommend *The Art of Styling Sentences*, 5th Edition, by Ann Longknife and K. D. Sullivan. For additional discussion of punctuation conventions, I recommend reading *Chicago*, Section 6, "Punctuation." Many of my students have benefitted from reading this particular chapter.

TABLE 4.1

The ways leptons interact with atoms in the literary universe, the chemical formulas of sentences. Fragments must be used with caution. Abbreviations for word classes and phrases can be reviewed in Tables 1.1 and 2.2, respectively. While these patterns are the ones I see most often in writing, this table is not an exhaustive list, so feel free to tinker, combine, and iterate. (Try replacing any "S V" pattern with a higher valency pattern from Table 3.1, for example.) Pay close attention to where punctuation and spaces go.

SENTENCE PATTERN	CHEMICAL FORMULA
Zero Independent Clauses (Fragment)	**S.** **V.** **Mod.**
with coordinating conjunctions	cc S. cc V. cc mod.
with subordinating conjunctions	sc S V. sc mod.
with another fragment	S cc S. S, S. S—S. S . . . S. S: S. S (S). V cc V. V, V. V—V. V . . . V. V: V. V (V). Mod cc mod. Mod, mod. Mod—mod. Mod . . . mod. Mod: mod. Mod (mod).

in a series of fragments	S, S, cc S. S, S, S. S—S—S. S . . . S . . . S. V, V, cc V. V, V, V. (etc.)
as a question (an interrogative fragment using a **wh-word: who, whose, whom, what, which, when, where, why,** or **how,** plus their **-ever** counterparts)	[*wh*-word]? [Who/whose/whom/what/which] S? [When/where/why/how] V? [*wh*-word] mod?
One Independent Clause (Simple Sentence)	**S V.**
with an introductory coordinating conjunction	cc S V.
with a modifier	Mod, S V. Mod—S V. Mod . . . S V. Mod: S V. S, mod, V. S—mod—V. S (mod) V. S V, mod. S V—mod. S V . . . mod. S V: mod. S V (mod).
with another independent clause as a peripheral element	S—S V—V. S (S V) V. S V (S V).

(*Continued*)

TABLE 4.1 (*Continued*)

SENTENCE PATTERN	CHEMICAL FORMULA
with inversions	V S. A S V. A V S. SP S V. SP V S. DO S V. OC S V DO.
as a question (an interrogative clause)	S V? p.aux S V? m.aux S V? [*wh*-word] p.aux S V? [*wh*-word] m.aux S V?
with a direct quotation and dialogue tags like *she said*	S V, "Quotation." "Quotation," S V, "continued quotation." "Quotation," S V. "Quotation," V S.
with a direct quotation with dialogue tags that skips source material	S V, "Quotation that skips material . . . and continues quotation from the same sentence." S V, "Quotation that skips material. . . . Continued quotation from a different sentence of the same source."
with a direct quotation that forms part of the surrounding syntax (no comma)	"Integrated quotation as subject" V. S "integrated quotation as verb." S V "integrated quotation as direct object." (etc.)
with a quotation, a pause, and commentary about the quotation	S V—"Quotation." S—"Quotation"—V. "Quotation"—S V. S V . . . "Quotation." "Quotation" . . . S V. S V: "Quotation." "Quotation": S V.

with a compound subject	S cc S V. S, S, cc S V. S, S, S, cc S V.
with a compound predicate	S V cc V. S V, V, cc V. S V, V, V, cc V.
with compound modifiers	Mod, mod, cc mod, S V. Mod, mod, cc mod—S V. Mod, mod, cc mod . . . S V. Mod, mod, cc mod: S V. S V, mod, mod, cc mod. S V—mod, mod, cc mod. S V: mod, mod, cc mod. S V (mod, mod, cc mod).
Two or More Coordinated Independent Clauses (Compound Sentence)	**S V, cc S V.** **S V, S V, cc S V.** **S V, S V, S V, cc S V.** **(etc.)**
when both clauses are very short	S V cc S V.
with **asyndeton,** a literary device where a series has multiple commas but no coordinating conjunctions (quickens pacing)	S V, S V, S V.
with **polysyndeton,** a literary device where a series has multiple coordinating conjunctions but no commas (slows pacing)	S V cc S V cc S V.
with semicolon(s)	S V; S V. S V; adv, S V. S V; S V; S V. S V; S V, cc S V. S V, cc S V; S V. S V SP; S, SP. S V DO; S, DO.

(*Continued*)

TABLE 4.1 (*Continued*)

SENTENCE PATTERN	CHEMICAL FORMULA
with a correlative conjunction	**If** S V, **then** S V. **If not** mod, **at least** S V. **The more** S V, **the more** S V. **Either** S V **or** S V. **Neither** S V **nor** S V. **Not only** S V **but also** S V. **As** mod **as** mod, S V. **As** mod **as** NP, S V. S V **as** mod **as** mod. S V **as** mod **as** NP. **Just as** S V, **so too** S V.
One Independent Clause + Dependent Clause(s) (Complex Sentence #1)	S V sc S V. sc S V, S V.
with two coordinated dependent clauses	S V sc S V cc sc S V. sc S V cc sc S V, S V.
in a series with asyndeton	sc S V, sc S V, sc S V, S V. If mod, if mod, if mod, then S V. When S V, when S V, when S V, S V.
One Independent Clause + Nested Dependent Clause(s) (Complex Sentence #2)	**[Dependent clause as S] V.** **S V [dependent clause as A, SP, or DO].** **S V IO [dependent clause as DO].** **S V DO [dependent clause as OC].** **S V DO [dependent clause as A].**
where the dependent clause is a modifier	Replace "mod" with [dependent clause as mod] in previous formulas
where the dependent clause has a ***wh*-word**	Replace "*wh*-word" with [dependent clause with *wh*-word] in previous formulas

where the direct object is a ***that*-clause** (*that* + S + V) (See the appendix for more about using the word *that* as a specific type of Function word called an **complementizer**.)	S V that S V.
where the complementizer *that* has been dropped from a ***that*-clause** (*I said [that] I would call.*)	S V ~~that~~ S V.
where the direct object is a series of ***that*-clauses**	S V that S V, that S V, that S V.
Two Coordinated Independent Clauses + Dependent Clause(s) (Compound-Complex Sentence)	**Mix and match!**
Common Errors (per *Chicago*)	**Numerous!**
no period (it happens to the best of us)	S V **Fix: S V.**
comma placed between subject and verb	S, V. **Fix: S V.**
comma placed between verb and direct object	S V, DO. **Fix: S V DO.**
second comma missing around an Appositive Phrase	S, App V. **Fix: S, App, V.**
serial comma missing before coordinating conjunction in a series	S, S cc S V. **Fix: S, S, cc S V.**
run-on sentence	S V S V. **Fix: S V. S V.** **Or: S V; S V.** **Or: S V, cc S V.**

(*Continued*)

TABLE 4.1 (*Continued*)

SENTENCE PATTERN	CHEMICAL FORMULA
comma splice, a comma between two independent clauses* *When clauses are repetitive and short, a comma splice is OK. Comma splices are most often found in fiction and in less formal dialogue or interior discourse.	S V, S V. **Fix: S V, cc S V.** **Or: S V; S V.**
compound predicate problem, an extra comma in a compound predicate sentence	S V, cc V. **Fix: S V cc V.** **Or: S V, cc S V.**
comma placed before restrictive relative clause	S V, that S V. **Fix: S V that S V.**
no comma before nonrestrictive relative clause	S V which S V. **Fix: S V, which S V.**
a hyphen or an en dash used instead of an em dash (in *Chicago* style)	S V - S V. (hyphen with spaces) S V-S V. (hyphen without spaces) S V – SV. (en dash with spaces) S V–SV. (en dash without spaces) **Fix: S V—S V. (em dash without spaces)**
hyphen used between series of page numbers	pp. 34-93 **Fix: pp. 34–93**
spaces missing between periods in an ellipsis	S V ... S V. (three dots) "S V.... S V." (four dots in certain quoted material) **Fix: S V . . . S V. (three dots)** **And: "S V. . . . S V." (four dots, no space before first period)**

final quotation mark placed before punctuation	"Quotation". "Quotation", S V. **Fix: "Quotation."** **And: "Quotation," S V.**
close parenthesis mark placed before punctuation	S V (mod.) **Fix: S V (mod).**

Note: Recall that a **modifier (mod)** is any part of speech that describes another word. Sometimes it is one word (possessive noun, adjective, adverb, determiner, pronoun, participle, infinitive, gerund, or appositive) or a phrase (Adjective Phrase, Adverbial Phrase, Prepositional Phrase, Verbal Phrase, Appositive Phrase, or Absolute Phrase).

EXERCISE 1

How do your favorite authors use the punctuation marks covered in this lesson? When you read the words out loud, can you hear, from how they've punctuated the sentence, where they want to pause, tighten, and link?

EXERCISE 2

Take the sentence formulas from Table 4.1 and create your own fragment or sentence according to the pattern, including common errors and their fixes. This is an exhausting but supremely instructive exercise for students who struggle with punctuation.

EXERCISE 3

Which sentence formulas do you like best? Identify your sentence formulas in one paragraph of your writing. Are they varied enough?

LESSON V

REPETITION, VARIATION

Super-Sentence Level

Every sentence has a rhythm of its own, which is also part of
the rhythm of the whole piece. Rhythm is what keeps the
song going, the horse galloping, the story moving.

—Ursula Le Guin, *Steering the Craft:*
A Twenty-First-Century Guide to Sailing the Sea of Story

IT'S FRIDAY NIGHT NOW, I'm here typing on my laptop
in my—*my*—library chair, and my boys are asleep in their beds. In a
few days, this will all be over. The end is in sight; I will soon put this
madness behind me Halloween night. In fact, I have a new sleep test
scheduled for Monday evening at a sleep clinic.

We now return to sound and movement, but this time we also notice silence and stillness, along with the sudden break in rhythm. The
brief lull between heartbeats in your pulse doesn't mean you're dead;
it means you're alive. We want to bring the same living pulse to our
proper, complete science sentences in what I call the **super-sentence**:
great sentences you want the reader to particularly remember.

In Lesson I, we studied sound and movement vibrating at the word
level, including syllables broken down into stressed and unstressed syllables (lexical stress). We also explored how the sound of the word can
be deeply connected to the meaning of a word (sound symbolism) and
that the meaning of the word may even be connected to the sound of

the word (the bouba/kiki effect). In Lesson II, we studied multi-word phrase patterns. In Lesson III, we studied syntax patterns. In Lesson IV, we amplified and clarified those patterns with more writing conventions, like punctuation, which mostly show us when to pause, which words link to other words, and what words are more important than others in a sentence.

And now, in Lesson V, we'll study repetitive stress patterns and variations in sentences, what is called **prosodic stress**. Prosodic stress, or sentence stress, builds on the natural lexical stress of words; on top of that, we can arrange whole phrases and clauses to emphasize important words and de-emphasize boring words. Mastering these wavelike rhythms and using occasional variations for emphasis (like taking a journey on the ocean with its rhythmic waves and occasionally spotting a dolphin or turtle or shark) helps readers focus on the most important parts of the sentence. And remember them.

Rhythm, in general, is the measured flow of words, phrases, and clauses within a sentence. The word is derived from the Greek *rhythmos*, measured motion. Rhythm in writing is molded by the patterns of stressed and unstressed syllables. This natural or default emphasis of words in a sentence is called *phrasal stress*, as opposed to purposeful emphasis to bring attention to a word or phrase, which is called *contrastive stress*. I'll call them **rhythm** and **emphasis**, respectively. We've already covered various options for emphasizing words in Lesson IV, but it's worth remembering that careful use of emphasis creates needed variation in the rhythm of your sentences.

Why would poetry-like rhythms ever be discussed in a book on science writing? In Subatomic Writing, writing that matters helps readers remember science information by tapping into humanity's love of repetition—and variation. For the last four years, I've asked students to share their favorite science sentences from their favorite science writing pieces. We analyzed the prosodic stress of each sentence. More often than not, science sentences that my students love contain what I'll call an **A Meter** and sometimes a **B Meter**, but the sentence doesn't have to fit perfectly into those rhythms. Sometimes, both rhythms are broken to bring emphasis to a certain concept within the sentence. Nevertheless, favorite sentences possess memorable, identifiable prosodic stress. The rest of this lesson will teach you the skill of scansion and describe how the "weak force" of the W^+ and W^- bosons is nevertheless a transformative power that stabilizes an atom, just as repetition and variation transmute a sentence into something more for readers.

SCANSION, PART I

Scansion is the writerly tool to help you identify rhythm and emphasis in a sentence. Consider this sentence from one of my favorite authors, Bill Bryson, from his science-writing masterpiece, *A Short History of Nearly Everything*:

> It is a slightly arresting notion that if you were to pick
> yourself apart with tweezers, one atom at a time, you would
> produce a mound of fine atomic dust, none of which had
> ever been alive but all of which had once been you.

This is a fabulous, funny, informational, philosophical sentence already, all on its own. It's perfect for Bryson's primary audience, in this case the public (Max loves this book even more than I do). But there's more! If I were to "poemize" this sentence, I'd break it into lines like this (yours might look a little different):

> It is a slightly arresting notion
> that if you were to pick yourself apart
> with tweezers, one atom at a time,
>   you would produce a mound of fine atomic dust,
>   none of which had ever been alive
> but all of which had once been you.

Bryson's sentence is now a fine stanza of free-verse poetry, which doesn't have to rhyme and isn't required to have a set number of syllable patterns per line (meter). In fact, in prose—including most forms of professional science writing—we don't aim for rhyme or regular meter at all, or if we do, it should be subtle and infrequent. But free verse still follows the rise and fall of human speech, which we will see in a moment. We love this natural flow of speech, even in writing, because we can still hear it in our heads. The wavelike rhythms carry us forward.

Notice, too, pinging back to Lesson IV, how Bryson has organized his phrases, clauses, and pauses with punctuation. Even though it's a long sentence—45 words, many of them Germanic Function words—we don't run out of breath when we read it. Often—but not always—you'll find punctuation where a new phrase or clause starts. A **run-on line** is a poetic line that doesn't have a natural speech pause at the end of it, and this kind of poetic line allows the sense to flow right into the next line. I've poemized line two of Bryson's sentence as a run-on line, for example.

But there's more! Bryson has written his prose-poem using the secret of repetitive stress patterns—but not perfectly regular patterns. This is what we want in science writing, in sentences you want the reader to especially remember: rhythm, but not too much rhythm.

In Microsoft Word—and in other word processors—you can use the Equation feature to add various dots and dashes to math and physics equations, like the waveform functions of particles (Insert menu→ "Equation"→"Accent" dropdown menu).

But you can also use the Equation feature to scan poetry, bringing a visual element to analyze an auditory phenomenon. Marking rhythm in writing is called scansion.

I think scanning is faster and more fun with paper and pencil than in Word, but I allow my students to scan the prosodic stress of sentences however they like. Whatever medium you choose, the process of scansion is the same.

First, you mark stressed and unstressed syllables with dashes and dots, or /'s and *u*'s—however you like. Use a W⁺ and a W⁻, if you like. Search for the word online at www.m-w.com (*Merriam-Webster*) to determine how the word is broken into syllables, and which syllables have primary and secondary stresses. Unstressed syllables have no extra accent mark before them; primary stresses have an accent at the top of the syllable, secondary stresses at the bottom of the syllable. This is *lexical stress* from Lesson I coming back to form the base of prosodic stress. Then, group those syllables into **feet**: one **foot** comprises two or more syllables that form a basic repeated pattern of stressed and unstressed syllables. The *iamb* is the most common kind of foot in the English language. The iamb's rhythm sounds like this: *da DUM.*

Next, identify the **meter**, the number of feet you count in one poemized line. Meter is the kind of rhythm we can clap to, where the beat (the pattern of stressed syllables) is arranged in equal intervals. If you put five *da DUM*s in a line, you'll have *iambic pentameter*, meaning five iambs, ten syllables total in the line. This is the rhythm Shakespeare used most, the rhythm we naturally use in speech.

Finally, note variations from that established pattern that draw the reader's attention. In Table 5.1, we see the pattern names of common rhythms we hear all the time around us. **Iambs** are related to **anapests**, and the rhythmic *opposite* of iambs and anapests are **trochees** and **dactyls**. The effect of moving from iambs to "opposite" trochees and back again is like spinning one way with your partner during a Viennese Waltz, reversing the spin, then reversing again to continue with a natural turn around the ballroom floor. This is moving from A Meter to B Meter and back again to A Meter.

Consider the pattern of Bryson's words, at least the way I say them out loud, broken down into stressed syllables (—) and unstressed syllables (·):

. — . — . . — . — .

. — . — . — . — . —

. — . . — . — . —

   — . . — . — . — . —

    — . — . — . — . —

. — . — . — . —

Putting it together, we get our poetic notation:

    . — . ——— . . — . — . .
It is a slightly arresting notion
    . — . ——— . ——— . — . ——
that if you were to pick yourself apart
    . ——— . . — . — . ———
with tweezers, one atom at a time,
    ——— . . ——— . ——— . — .— . ———
you would produce a mound of fine atomic dust,
    ——— . ——— . — . ——— .——
none of which had ever been alive
    . — . ——— . ——— . . ———
but all of which had once been you.

The way I read it, Bryson's sentence has an A Meter and a B Meter: first comes *iambic* meter, a string of iambs put together. Line two of Bryson's sentence has exactly five iambs in a row (ten syllables): perfect iambic pentameter. When I ask my students to write Shakespearean science sonnets, this is the rhythm each line has.

Then comes a switch to a *trochaic* type of rhythm (but not exactly) starting in line four, with a turn back to iambic meter in the last line. We'd never claim this is a fixed verse, and none of the other lines have iambic pentameter, but Bryson has used rhythm to his advantage. He builds the pattern, breaks the pattern, and returns to the original pattern. All in one sentence.

Did Bryson write this way on purpose? I haven't asked him, but no, I seriously doubt it. Great writers often employ rhythm because "it sounds good." But this is hardly useful writerly advice. Why does it sound good? Because the writer is paying attention to the subatomic

TABLE 5.1

Simple scansion: The most common rhythms in speech and poetry.

NAME OF FOOT	NAME OF METER	SYMBOLS	EXAMPLE
Iamb	Iambic	. —	. — *the sun* . — *around*
Anapest	Anapestic	.. —	. . — *by the lab* . . — *intervene*
Trochee	Trochaic	— .	— . *went to* — . *enter*
Dactyl	Dactylic	— ..	— . . *color of* — . . *nucleus*
Pyrrhic	Pyrrhic	..	. . *of a*
Spondee	Spondaic	— —	— — *true blue*
Monosyllable	Monosyllabic	—	— *truth*

levels of writing with repetition and variation, and this includes prosodic stress. If you read a lot of great literature, or study music, you can more easily adopt these rhythms naturally in your writing, but it can help to study prosodic stress directly through scansion.

W⁺ AND W⁻ BOSONS: A CONTINUATION OF THE WEAK FORCE

That first night, I chose the W bosons to represent the ebb and flow of prosodic stress because I figured "W⁺ and W⁻" would make it easier to remember the point of this lesson, repetition and variation. But the more I study W bosons, the more I like them as representative forces of this lesson, and here's why: prosodic stress is a continuation of punctuation, emphasis, and spacing, just as the W interactions join Z as part of the same fundamental force, the weak force.

In fact, Z and W themselves interact: one experiment at CERN—ATLAS—measured two W bosons, each of them radiating off a quark in a proton and fusing together to make a Z boson. The same can be said for how words affect rhythm and rhythm affects punctuation in writing.

Weak as these bosons are, at least compared to the strong force, the weak force is still stronger than gravity. Curiously, all three weak-force bosons have a significant amount of mass themselves—that is, they're substantive in their own right, in our metaphor. The W bosons are 80 times more massive than a proton, Z bosons over 90 times. W^+ even has the power to decay into different combinations of leptons and neutrinos—or even special mesons called *D mesons*.

Given the power of sound and motion that punctuation and prosodic stress lend writing, perhaps this is why poetry, emojis like ༄("^˘ŏŏŏ̆^")ψ, or even ". . ." in a text message can convey a message beyond formal sentence constructions. When we study poetry in my science writing classes—my favorite modern poets are Major Jackson, Aimee Nezhukumatathil, Ross Gay, Tracy K. Smith, and, naturally, all poets who graduated from my alma mater, Bennington College—my students often struggle to find "the meaning" of a poem. I tell them to let go a little and focus more on what the words and rhythms cause them to think and feel. While the goal of science writing is to communicate information, this goal shouldn't stop you from writing in a way that helps your readers stay with your message—and your message stay with your readers.

In fact, the W bosons increase the stability and durability of atoms, in what is called the **transmutation** of quark flavors. While this makes bosons sound like a heavenly dessert, it means that W bosons are transformational. They take the less stable quarks—Charm, Strange, Top, and Bottom—and move them toward the Up and Down side of the family. A W$^+$ turns Up to Down, Charm to Strange, Top to Bottom. A W$^-$ turns Down to Up, Strange to Up, and Bottom to Charm. As the W bosons turn bloated quarks into more common matter, other leptons and even quarks spark off in pairs. Perhaps, by analogy, rhythm in writing is transformational and adds this high-energy spark to your words. If you are writing a sentence, and you have to choose between a perfectly concise, no-messing-around sentence versus a rhythmic sentence that perhaps contains a few extra words to make it rhythmic, I say go for the "super-sentence." *Quantum* does mean the minimal possible amount of something involved in a physical interaction, but sometimes too little communication can create confusion and boredom.

Remember, in Lesson III, how two protons collide in the Sun, and in one of the protons, the Up quark changes to Down so the proton becomes a neutron? This is part of W$^+$'s job, transmuting Up to Down and thus allowing a deuterium nucleus to form in the Sun's fusion process. When we write with rhythm in mind, sometimes we have to switch our words around—exchanging a stressed Lexical word for an unstressed Function word—to get the combinations of metrical feet that sound better.

Conversely, when a W$^-$ turns Down to Up, this is specifically called **beta decay** (remember, in Subatomic Writing, decay is a good thing). Sometimes atoms have too many neutrons, and this doesn't make an atom happy. (Similarly, when a sentence has too many boring Function words, it needs more common Lexical words or greater concision.) Beta decay, facilitated by W$^-$, turns the neutron (ddu) into a proton (uud), releasing an extremely energetic electron (called a **beta particle**, often seen with the Greek letter B, called beta: β) and an antineutrino. This radioactive transformation moves the nucleus up an element on the periodic table.

Sometimes, beta decay can give the promoted atom so much energy that it may even emit a DNA-penetrating photon (a **gamma ray**) in the process. For example, here's an unhappy boron-12 atom decaying into an excited carbon (represented as C*)—along with the high-energy electron and its antineutrino. Then, the excited carbon releases a photon and becomes stable:

$$^{12}_{5}B \rightarrow \, ^{12}_{6}C^* + \, ^{0}_{-1}e + \bar{v}_e$$

$$^{12}_{6}C^* \rightarrow \, ^{12}_{6}C + \gamma$$

Thus is boron transmuted to carbon through glorious, stabilizing beta decay, releasing energy and light. Beta-decay writing, if you're lucky, might also energize and change a reader, slightly, forever. Rhythm is powerful that way. I've listed many "class favorites," along with my own favorite sentences, in Exercise 2 at the end of this lesson.

There are Feynman diagrams for each of these interactions, but we'd better get back to scansion.

SCANSION, PART II

As we scanned sentence after sentence, I noticed that some of the feet we were seeing didn't match any of the metrical feet I'd learned in high school English classes. I saw in science prose the unusual (for strict poetry) pattern of unstressed-stressed-unstressed, either as a foot to break up a rhythm that was becoming too regular or as a repeated pattern itself. This foot also has a name: the *amphibrach*, with its opposite, the *cretic* foot. Table 5.2 shows the second wave of scansion patterns that you might find or use yourself as you transmute your writing with prosodic stress.

I usually recommend that students stick with the rhythm patterns in Table 5.1 until they're comfortable with the basics of scansion. You quickly discover that one sentence can be read several different ways. Even though lexical stress stays the same for everyone (this is the primary and secondary stress of a word as printed in the dictionary, and is usually nonnegotiable), prosodic stress can change depending on how someone reads the line. Sometimes "like calls to like," and stronger patterns (often based on lexical stress) affect the syllables around them. The trick to diminishing uncertainty in rhythm is to ask a friend to read aloud a "radiant" sentence you've crafted, and see if your friend speaks the sentence in the same way you intended it to be heard.

TABLE 5.2

Advanced scansion: Less common trisyllabic feet found
in poetry and prose.

NAME OF FOOT	SYMBOLS	EXAMPLE
Tribrach	. . .	*such as a*
Molossus	— — —	*milk maid woes*
Amphibrach	. — .	*the morning*
Cretic	— . —	*Tom the cat*
Bacchius	. — —	*when night falls*
Antibacchius	— — .	*hard choices*

Note: If you're brave and want to try *tetrasyllable* feet, look up "Foot (prosody)" on Wikipedia and scroll down to the table titled "Tetrasyllables."

To help you with trickier elements of scansion, what follows are some additional poetry terms regarding repetition and variation. In providing more poetry jargon, I hope to give you a secure foothold in what might be new terrain for you. I personally like having names for things, if you can't tell. Finding the name of a phenomenon I've been wondering about for a long time gives me joy—and some measure of comfort that other people have paused to consider and give a name to the same phenomenon. All of the following terms can be applied to great science writing in some small way.

Here are some additional terms to help you with missing or extra stresses or metrical feet as you scan:

Grammatical pause (also called a *caesura*): A pause in the reading of a line by some kind of punctuation mark (commas, periods, dashes, colons).

Rhetorical pause: A natural pause caused by a phrase or syntax.

Metrical pause: A pause that fills the place where you'd otherwise expect to see a stressed syllable. These kinds of pauses affect scansion, unlike *rhetorical pause* and *grammatical pause*.

Expected rhythm: The expectation set up by the repetition of feet.

Enjambment: A line that "runs over" to the next; a line with incomplete syntax that ends and continues to the next line. Used to create tension, enjambment is an ancient technique: the Greek poet Homer used enjambment in the *Iliad* and the *Odyssey*.

End-stopped line: A line that ends with some kind of punctuation or a natural speech pause.

Catalexis: A line that is missing a syllable at the end, a syllable you're expecting to see because of the expected rhythm. Catalectic endings were common in ancient Greek drama. A line missing two syllables is called *brachycatalectic*.

Headlessness (also *acephalous*): A line that is missing the first syllable so that it no longer matches the expected rhythm. Headlessness is common in anapestic meter, like in limericks.

"Limericks are the best," I heard a slurred voice say from right above my head. "I got lots of them."

It was Dyna, my empty bottle of Scotch in her claws, her horrid demon face leaning over the edge, right above my Shakespeare shelf. She belched.

I leapt out of my chair and faced her. "This is most unprofessional," I snapped. "I cannot concentrate with an intoxicated demon around. Besides, limericks are the least professional kind of fixed poetic form in the entire literary universe."

"Just tryin' to help," she said, hurt. "Also, I think you mean *intoxicating* demon. I'm amazing." She stood and walked unsteadily along the top of the bookshelves. Her same red Hawaiian shirt was crumpled and had burn marks on it. "Gutenberg thought so, too. I helped him base his printing press off a wine press. True story. He was from a part of Germany that produced lots of wine. He knew *aaaalllllll* about wine presses. Man, what a great time we—"

"Watch out!"

She knocked over my constellation globe, and it cracked in half.

"Oopsy," she said, turning back to me. Then she sneezed violently three times, emitting several black-blue sparks. Recovering, she looked around her. "Damn, girl," she said, tottering precariously close to the edge, glass bottle still in hand. "Don't you ever dust up here?"

"Last month," I lied. "Can you leave now?"

She sneezed again and sat down, feet dangling over the shelf again. "I gotta tell you my limerick first," she said. "It's the best, lots of anapests." She cleared her throat, half-hiccuped, and recited:

> There was a young maid who said, "Why
> Can't I look in my ear with my eye?
> If I put my mind to it,
> I'm sure I can do it
> You never can tell till you try."

"Need some more 'xamples?" she said, looking pleased with herself. "I got tons. What 'bout Little Willie?"

"Little Willie?" I asked, immediately regretting my question.

"Classic poetry!" she cried gleefully. "Trochees with some catalexis." She swayed back and forth as she recited:

> Willie saw some *dyna*mite,
> Couldn't understand it quite.

Curiosity never pays.

It rained Willie seven days.

I shut my laptop and spoke with my nicest, most soothing mom-voice. "OK, Dyna, thank you. I got it. Very helpful. We've got all we need now to teach rhythm for Subatomic Writing," I said. "So thanks for the poems. I'll see you later, OK?"

Dyna set the bottle down and held back a sneeze by pinching her nose. She stated nasally, "You didn't like them? Your granddaddy loved that old joke book. First printed in 1948. *Your Own Joke Book.* What a classic. It's here on your shelf, you know." She patted the bookshelf. "You really should read these more often. You know, for fun and enjoyment. Enjambment. Enjoybment."

"I'll keep that in mind, thanks."

"You know, you an' me, JZ, we're just a couple of weak bosons. We got *substance*, you know what I mean?"

"Are you saying we're massive?"

"No, no, it's a compliment, a *compliment*, JZ. You an' me, we just interact with life, you know? I'm the positive one, you're the negative one, we try an' keep the whole world together . . ." As Dyna threw her arms out wide, I watched helplessly as the Scotch bottle rolled off the end of the bookshelf, whiffed through six feet of air, and shattered on my wooden floor.

"I love gravity so much," Dyna sniffed. "What a joy. What an enjoyjambment, of space and time together, accelerating matter from one spacetimeline right into the next. So sad you mortals don't understand it yet." She squinched her eyes shut and held her breath— presumably to hold in a sneeze. After a moment, she relaxed and leaned over the edge of the bookshelf.

"You know you should recycle that, right?" she said, pointing to the glass mess. Before I could respond, she sneezed again, so violently that she disappeared in an explosion of sparks.

Halloween night can't come soon enough this year.

Fortunately, we're almost done for tonight, despite demonic interruption.

SCIENCE POET MODELS

While great science writers transmute their sentences into balanced, energetic sentences, you can absorb rhythm by interacting with poetry on its own, even poetry from other scientists and science writers, past or present. The stereotype of "science versus English" is old and boring and counterproductive. We need science writing—all kinds of science writing—to engage and illuminate readers better through the power of prosodic stress.

In 2016, the literary magazine *Kenyon Review* created a special issue on "The Poetics of Science." I was delighted to be a part of it, and I am still friends with the poet-scientists I met during the production of the issue. Sergei Lobanov-Rostovsky, the associate editor, was fantastic to work with. In the introduction of the print issue, he postulates, "How does science inspire the literary imagination? Can science writing be a literary art?" He also hosted an online conversation between contributors: https://kenyonreview.org/2016/09/poetics-of-science/.

I've also had the pleasure of exchanging emails with Dava Sobel, one of my favorite science writers. She is now the editor of the "Meter" column for *Scientific American*, which is dedicated to science poetry. She told me she loves diagramming; in fact, she took an elective seminar in syntax, and their textbook was *The Hobbit*. "Each student," she said, "had to diagram one chapter. It was revelatory." Science, grammar, poetry: these were never isolated disciplines. Dyna's ancient Greek influence is perhaps a reminder to us to stop thinking of them as separate entities. We might progress faster and further in science and public education if we do.

Sobel also directed me to a brilliant book called *A Sonnet to Science: Scientists and Their Poetry* by Sam Illingworth. From it, I learned that James Clerk Maxwell wrote at least one poem a year from the time he was a teenager until he died—except during two periods: when he was a student writing "On the Theory of Rolling Curves" and "On the Equilibrium of Elastic Solids" and when he mathematically proved Michael Faraday's "lines of force" by writing "On Physical Lines of Force." During this same time, he also demonstrated that light was, in fact, an electromagnetic wave in "A Dynamical Theory of the Electromagnetic Field." When he wasn't breaking new ground in physics, he was consistently writing poetry.

Salman (Sal) Khan, an American educator and founder of the nonprofit Khan Academy, is one of my major heroes. Max didn't skip a beat during COVID-19 shutdowns because I assigned him Khan Academy schoolwork, and I've used Khan Academy myself to resurrect my math skills. In his book, *The One World Schoolhouse: Education Reimagined*, Khan writes, "The truth is that anything significant that happens in math, science, or engineering is the result of heightened intuition and creativity. This is art by another name, and it's something that tests are not very good at identifying or measuring." The influence of prosodic stress in science writing similarly cannot be measured, but its influence is unquestionably felt by readers.

EXERCISE 1

Select two or three of your favorite sentences from literature and identify the scansion patterns. First, read the sentence out loud, a few times if you need to. Mark stressed and unstressed syllables (any way you like, paper and pencil if possible; colors help too). Next, identify feet using Tables 5.1 and 5.2. See if you can detect an A Meter and a B Meter. Finally, if there is any regular meter, ask: Does the writer break the rhythm to bring particular emphasis to one word or phrase in the sentence?

EXERCISE 2

Scan any or all of the following quotations. See if a friend pronounces the same sentence differently than you do.

Physics is really nothing more than a search for ultimate simplicity, but so far all we have is a kind of elegant messiness.
—Bill Bryson, *A Short History of Nearly Everything*

Nobody knows for sure how many holes you have in your skin, but you are pretty seriously perforated.
—Bill Bryson, *The Body*

The transparent ice along the lake's edge is filled with bubbles of air, trapped inside like the sustained notes of a soprano.
—Terry Tempest Williams, *Refuge*

Inside the frame, through hundreds of revolving seasons, there is only that solo tree, its fissured bark spiraling upward into early middle age, growing at the speed of wood.
—Richard Powers, *The Overstory*

We are often like rivers: careless and forceful, timid and dangerous, lucid and muddied, eddying, gleaming, still.
—Gretel Ehrlich, *The Solace of Open Spaces*

Matter changes hands, atoms flow in and out, molecules pivot, proteins stitch together, mitochondria send out their oxidative dictates; we begin as a microscopic electrical swarm.
—Anthony Doerr, *All the Light We Cannot See*

In every outthrust headland, in every curving beach, in every grain of sand there is the story of the earth.
—Rachel Carson, *Holiday*

Chaos breaks across the lines that separate scientific disciplines.

    —James Gleick, *Chaos: The Amazing Science of the Unpredictable*

Six great blue herons, hoping for an easy dinner, have gathered, wing to wing, on the bank, like students waiting in line in a cafeteria.

    —Elizabeth Kolbert, *Under a White Sky: The Nature of the Future*

If my decomposing carcass helps nourish the roots of a juniper tree or the wings of a vulture—that is immortality enough for me.

    —Edward Abbey, *Desert Solitaire*

> we're anything brighter than even the sun
> (we're everything greater
> than books
> might mean)
> we're everyanything more than believe
> (with a spin
> leap
> alive we're alive)
> we're wonderful one times one.

    —Edward Estlin Cummings (1894–1962), last stanza of the poem "if everything happens that can't be done." e e cummings, as he wrote his name, was known for actively rejecting word class, syntax, and uppercase letters.

In certain latitudes, there comes a span of time approaching and following the summer solstice, some weeks in all, when the twilights turn long and blue.

    —Joan Didion, *Blue Nights*

The scientific vocabulary is the bridge by which we enter the land, not the wall that keeps us out.

—Isaac Asimov, *Words of Science and the History Behind Them*

Among the earliest forms of human self-awareness was the awareness of being meat.

—David Quammen, *Monster of God*

I want and want and want but never allow myself to reach for what I truly want, leaving that want raging desperately beneath the surface of my skin.

—Roxane Gay, "What Fullness Is"

She was beautiful, but she was beautiful in the way a forest fire was beautiful: something to be admired from a distance, not up close.

—Neil Gaiman and Terry Pratchett, *Good Omens*

Who we are is inseparable not only from who we think we are but from who others think we are.

—Amy Ellis Nutt, *Becoming Nicole*

We were not designed rationally but are products of a convoluted history.

—Neil Shubin, *Your Inner Fish*

With the naked eye, I can see two light-years into the Andromeda Galaxy.

—Annie Dillard, "Seeing"

They use everything about the hog except the squeal.

—Upton Sinclair, *The Jungle*

There is a nothingness of temperature, a point on the body's mercury where our blood feels neither hot nor cold.

—Jill Christman, "The Sloth"

So the question arose in my mind that day: Did I choose to plant these potatoes, or did the potato make me do it? In fact, both statements are true.

—Michael Pollan, *The Botany of Desire: A Plant's-Eye View of the World*

The Universe is an incredibly hostile place for life.

—Phil Plait, *Death from the Skies: The Science Behind the End of the World*

Time is to clock as mind is to brain.

—Dava Sobel, *Longitude: The True Story of a Lone Genius Who Solved the Greatest Problem of His Time*

In the face of velocitized steel, even the strongest among us are mush.

—Mary Roach, *Grunt: The Curious Science of Humans at War*

Language, that most human invention, can enable what, in principle, should not be possible. It can allow all of us, even the congenitally blind, to see with another person's eyes.

—Oliver Sacks, *The Mind's Eye*

One does not see with the eyes; one sees with the brain, which has dozens of different systems for analyzing the input from the eyes.

—Oliver Sacks, *Hallucinations*

DOT-TO-DOT GAME

Paragraph Level

Pick a flower on Earth and you move the farthest star.

—Paul A. M. Dirac

SOMETIMES, when your scattered thoughts are out there floating on their own, the hardest act of creation in the literary universe is to draw those disparate ideas into a coherent clump of matter called the **paragraph**. Like drawing invisible lines between stars to make constellations, readers must draw lines between our facts and ideas from sentence to sentence within a paragraph. We can't do it for them, but if we organize the dots and leave clues about how one point relates to the next, they will be able to connect the dots and see the big picture we are attempting to communicate.

This dot-to-dot game at the paragraph level is the topic of our final lesson. We'll cover our last boson, the **graviton**, and we'll take a quick tour of pacing, coherence, and concision techniques at the paragraph level. We'll see these skills at work in speeches from two cosmic ray physicists, and we'll complete the final step from a carbon atom to a chain of carbon atoms: graphite.

THE GRAVITON

Gravity might get a bum rap in quantum physics for being the weakest of all four forces, but its power reaches through vast distances, pulling

individual atoms into coherent clouds of dust. That dust starts to co-
alesce and spin together, hotter and hotter, gathering more attractive
force, tighter and denser, gaining angular momentum—

And then a star bursts into being. This is how our Sun formed 4.6
billion years ago.

The lingering bits from the dust cloud (about 0.01 percent of the
original mass of dust) swirled around the Sun. Drawn into the groove
that the Sun's gravity carves through spacetime, the bits of dust co-
alesced into the planets of our Solar System.

Matter attached to one of these planets also clumped together, and
with the help of the other forces, gravity built rocks, plants, animals,
us. The phenomenon of gravity still governs our paths around the
Earth, the Sun, the Milky Way, the Local Group, the Virgo Cluster, the
Virgo Supercluster, Laniakea, and unimaginably long filaments in our
observable universe. And yet, no matter how small the matter, there is
still a weak attraction between all those particles.

In high school physics classes, we'd calculate the force of attraction between two students—I'm 100 percent sure we were the very first
teenagers to do this—by multiplying the masses of two students together, dividing by the square of the distance between their seats in
the classroom, and multiplying by Newton's gravitational constant, G.
We were always underwhelmed that the force of attraction was so small
a number, yet there was a kind of subconscious comfort in knowing
that just by existing, by being a mound of fine atomic dust in the cosmos, we were always connected to each other—as well as to our ultimate progenitors, the stars—by the sheer curvature of spacetime. In
other words, we're all warped, but our warped nature connects and
grounds us.

The **graviton** is the as-yet undiscovered exchange particle for the
gravitational force, the sixth boson as I see it. If/when we do find the
graviton, we know approximately what characteristics it might possess:
zero mass, neutral electric charge, and a "spin" of two. Higgs, a *scalar*
boson, has a spin of zero; the *gauge* bosons—gluons, photons, and Z

and W bosons—all have a spin of one. "Spin" for subatomic particles is an even worse analogy than color charge is for quarks. Google "particle physics spin" if you'd like a headache. It's Saturday night and I'm not touching that.

So we sort of know what the graviton might look like, and that's a good step. It's often the case that a particle is theorized and then found later, like many atomic elements, electron neutrinos, the Higgs boson, and many other particles, so there's hope. In fact, the recent discovery of gravitational waves is indirect evidence for gravitons. Henri Poincaré hypothesized gravitational waves in 1905; Albert Einstein predicted them in 1916; and in 2015 the Laser Interferometer Gravitational-Wave Observatory (LIGO) directly observed how the Hulse–Taylor binary pulsar created these waves. One thing builds on another. We build on each other's work. If you're out there, graviton, we will find you. By Dyna's wiry hairs, we will find you. Who knows, maybe there are two gravitons, like W^+ and W^-, that are working in conjunction to curve spacetime.

In the meantime, we know we at least experience the force of gravity, a long-range force that is 10^{39} times "weaker" than the strong force. And yet, gravity holds us and galaxies together. The same should be true of sentences, small or big, within our paragraphs. **Pacing** and **coherence** are two principles that make our writing more attractive.

PACING

Pacing is how slowly or quickly your narrative—that is, the story, the arguments, the information you want to share—unfolds through your sentences. Imagine you're a tour guide in Nicaragua driving a tour bus up Mombacho Volcano. If you drive too slowly from point to point, your tourists will become frustrated and bored. If you drive too quickly, they'll get frustrated and angry, having missed what they came to see (or they will fear for their lives: the road up the volcano slopes so steeply you can almost divide it by zero). Control the speed of your paragraph-tour using the following five pacing strategies.

Balance Summary and Scene

Let's review the difference between a summary and a scene. A **summary** skims through several potential scenes, giving an overview, explaining how things work, moving through time. Summaries are great when there's not much happening. Keep summaries short to increase pacing. It's like a tour guide driving past several mildly interesting trees or birds, slowing down a little to talk about them but not totally stopping the tour bus.

Generally, you want to present your readers with **scenes** rather than summary (although it depends on what kind of science writing you're doing). Summary is "telling"; scene is "showing." Readers like "showing" better. Any kind of science information can be "told" in summary or more carefully woven into a "shown" scene.

A scene stays in the moment, building tension and letting the story heat up. It's similar to stopping the bus so your tourists can watch the howler monkeys fight. You control how much time elapses before you move on. Description is important, but too much can slow down the pacing of the scene—think of the tour guide citing a slew of historical facts about monkeys while monkeys are throwing feces at the tourists. The tourists just need to stay in the moment, and the tour guide needs to shut up—and know when to drive on.

Action doesn't wander, it isn't focused on description, it doesn't linger on what people are thinking about the action. It's *action*. It moves. A set of events that quickly occurs one after the other, without lingering on transitions, increases the pacing, too. Strong verbs become your best friends in this case. In case you think formal science writing doesn't need any story at all, I invite you to reconsider the reader's needs. You can still build a story—an honest, accurate one, please, if you're writing science nonfiction—around abstract concepts by treating your nouns like actors, your verbs like actions, and your modifiers like the props and scenery of a play. As mentioned in Lesson I, let particles scatter, proteins build, and viruses die against the backdrop of your research.

Scene cuts—jumping to a new scene—can increase pacing if done well with minimal transitions. But don't lose the reader during the jump forward or backward. They'll get whiplash. I often recommend "time stamps" and "place stamps" to my students at the beginning of new paragraphs if they're jumping to a new time and place in a narrative—even within an abstract science narrative.

Build Suspense

As you drive up the volcano, you enter a cloud forest ecosystem, creating mystery and obscuring the view. Pacing actually increases when you don't immediately give the reader the resolution of a conflict. Cliffhangers at the end of novel chapters likewise draw readers right into the next chapter. Use uncertainty (which readers love and hate, but mostly love) to increase interest. Staying in a tense scene instead of rushing to a safer time or place also increases the pacing.

Shorten and Organize Paragraphs

When I see a paragraph that takes up a whole page, I cry. OK, I don't cry, but a little piece of me dies. OK, that doesn't happen either, but epic paragraphs feel like more of a chore than a pleasure to read. It slows down the pacing. Aim for about three to five sentences per paragraph, on average. Even the most stalwart of tourists needs a pee break once in a while.

Have a purpose for each of your paragraphs instead of wandering aimlessly on crumbling roads. The first sentence (the topic sentence, in more formal science writing) often tells the reader what to expect in the rest of the paragraph. The middle of the paragraph expounds on, adds to, qualifies, or proves the first sentence. The last sentence is the dramatic position of the paragraph if you'd like to be dramatic, the humorous position if you'd like to be funny, or the recap position if you'd like to emphasize the main point of the paragraph. It often looks forward, preparing readers for the topic sentence of the next paragraph.

Vary and Simplify Sentence Syntax

I'm not against the long sentence, but be aware that—especially in more journalistic pieces—your editor will probably prefer shorter sentences to long ones. Shorter sentences mean faster pacing. No one likes a tour guide who drones on and on.

If your sentence structure is complicated—if you have complex-compound sentences with lots of relative clauses, complements, nested PIG Phrases, etc.—your pacing slows down because the reader must work their way through your syntax. If this is the case, rephrase your sentences using simpler verb constructions (Table 2.1), phrase structures (Table 2.2), valency patterns (Table 3.1), and sentence formulas (Table 4.1) to increase your pacing. (Using items closer to the top of the tables will increase your pacing.) Active voice will likewise read faster than passive voice. There's nothing wrong with a tour guide slowing down to examine a particular plant, waxing eloquent and lyrical about the beauty of nature, or launching into a sermon about the conspicuous consumption of tourism, but be conscious of your audience's tolerance for such.

Revisit Word Choice

As a reminder from Lesson I, shorter, Germanic-based English words will read faster than longer Latin or Greek words; for example, *see* will read faster than *observe*. Concrete words will also be more accessible to the reader than abstract words. Let the physical senses—sight, hearing, taste, touch, smell—guide you in choosing vibrant Lexical words. Chopping adjectives and adverbs (which often "tell" the reader how to feel about a subject) is a general strategy but difficult to do if you're trying to describe a science phenomenon. Using words with metaphors baked into them (*chop*, *bake*, and *stew*, for example, are common cooking words you can apply elsewhere) and turning complex words into common words (Table 0.1) by using a thesaurus also helps your reader.

All writers, including editors, need to go through the painful process of cutting extraneous words in their pieces. It must be done (although give yourself as much freedom as you need in first drafts). Cut out the throat-clearing and false starts. You want to be a tour guide who speaks concisely and with confidence. What does your audience really need to know? What do you expect them to do with your information? Why should they care? Concision extends to dialogue: even if you've spent hours interviewing a scientist, cut out the dialogue that veers off-topic or restates what's already been said.

Concision is painful, no question. The "fluff" of your piece is like empty calories—your reader is quickly filled with them, but they're not very substantive. If you pass out snacks during your tour, don't let your reader leave a paragraph feeling bloated and unfulfilled.

You might need a few drafts to figure out what you want to write about—to outline the most logical stops of the tour and practice what you want to tell people. That's all part of the process. Once you (mostly) know what your story is, determine which parts of your story are scene-worthy and which can be summarized with description and facts. Ask yourself where your reader-tourists will want to linger and where they might get bored. If you've just seen the volcano craters, go see the sloth and the rare endemic salamander—*Bolitoglossa mombachoensis*—and leave the volcano-grown coffee for the finale. Pace your story well, and your tourists will leave informed and satisfied. If you've done your job well—well enough, because it doesn't need to be perfect to be great—they'll tell their friends to take your tour, too.

COHERENCE

In his book *The Sense of Style*, Harvard professor Steven Pinker addresses common writing problems. "Even if every sentence in a text is crisp, lucid, and well formed," Pinker writes, "a *succession* of them can feel choppy, disjointed, unfocused—in a word, incoherent." Your reader must not only understand your sentences but also be able to move through them, recreating the logic you used to write them.

Principles of Coherence

Sticking to the tour-guide metaphor, pretend you offer a separate bird-watching tour as you lead your tourists on foot around the longer Puma Trail at the top of Mombacho Volcano (pacing principles still apply). If you don't pay attention to where you're going and let your reader get too close to the snakes, the jaguar, or the edge of one of the four craters, you'll probably get bad reviews on TripAdvisor.com, if anyone makes it back down the volcano alive. Here are some key principles to connect your sentences together to create coherence:

Have a clear and consistent topic throughout the whole piece. As they say in journalism, don't bury the lede. Keep the main words of your topic as the subject (not always, but often). If you set out promising to show your tourists the blue-crowned motmot on your tour but you end up mostly talking about turkeys, your clients are going to be irritated.

Use paragraph breaks. Have mercy on your tourists and give them a chance to take a breath before plunging deeper into the cloud forest.

Clearly signal the logic between sentences. As you hike along the volcano, the path you choose, as the guide, starts becoming erratic. You climb over fallen trees when it's easier for everyone to go around. Without warning, you march down a path to the left that is significantly muddier than the one to the right. You disregard warning signs. Finally, you plunge right into thorn-filled foliage, and your followers just can't follow you anymore. You've lost their trust. They need a guide who will lead them along a logical path with identifiable checkpoints.

The same goes for the paragraph. Returning to *The Sense of Style*, Pinker built on the work of eighteenth-century philosopher David Hume and more modern linguists like Andrew Kehler to identify 15 coherence relationships. I recommend, if you particularly struggle with logic and coherence, that you buy the book and read the whole chapter ("Arcs of Coherence"). Here is my quick summary of Pinker's 15 logic paths between sentences, along with typical connecting words that signal the relationship:

1. **Similarity**: *and, similarly, likewise, too*
2. **Contrast**: *but, in contrast, on the other hand, alternatively*
3. **Elaboration**: *(colon), that is, in other words, which is to say, also, furthermore, in addition, notice that, which*
4. **Exemplification**: *for example, for instance, such as, including*
5. **Generalization**: *in general, more generally*
6. **Exception (generalization first)**: *however, on the other hand, then there is*
7. **Exception (exception first)**: *nonetheless, nevertheless, still*
8. **Sequence (before-and-after)**: *and, before, then*
9. **Sequence (after-and-before)**: *after, once, while, when*
10. **Result (cause-effect)**: *and, as a result, therefore, so*
11. **Explanation (effect-cause)**: *because, since, owing to*
12. **Violated expectation (preventer-effect)**: *but, while, however, nonetheless, yet*
13. **Failed prevention (effect-preventer)**: *despite, even though*
14. **Attribution**: *according to, stated that*
15. **Other, e.g., anticipation of a reader's reaction**: *yes, I know*

Use connectives—but use them carefully. Although you can see which connectives—also called **conjunctive adverbs**—go with each of the coherence relations, don't always use them. In fact, if you overuse them, you've ruined the reader's dot-to-dot game, placing too many signposts close together. Too much hand-holding feels patronizing to the reader. Too little, however, particularly on difficult terrain, leaves the reader feeling abandoned. "When in doubt," Pinker says, "connect."

Present old information in a sentence before you present new information. We learn by building on precepts we already know. Help frame new information for your reader by first referring to what they probably already know (something you assume your specific audience already knows or a concept you've introduced in an earlier sentence). Then connect the old information to new information, including jargon.

Use pronouns and adjectives to connect to the previous sentence. If Dyna, a character in your narrative, has already been introduced in a previous sentence, then connect the next sentence by using the personal pronoun *she* or the possessive determiner *her*. The definite article *the* implies that the reader already knows which thing you're talking about. Demonstrative pronouns and adjectives (*this, that, these, those*) and relativizers (*wh*-words) can also connect to the previous sentence. Just be careful that the antecedents of these words—particularly the personal pronoun *it* and the demonstrative adjectives (*this, that, these, those*)—are crystal clear. Pronouns should match the nearest preceding noun.

Repeat key words / use synonyms. Analyze any solid science essay and you'll find that the key terms are repeated throughout. Although Pinker cautions about the ways synonyms can derail coherence, a sprinkling of synonyms can keep the text fresh and still leave a topic trail for the reader to follow.

Are you going to think about all this information whenever you sit down to write a paragraph? No. I don't even think you should. But after the first draft or so, when you're revising, take a moment to check how well your sentences link together. If you don't feel your message is being relayed clearly, these are principles that can help you become a better tour guide at the paragraph level. Once you've left a connecting trail of dots in one paragraph, you can move to the next, then the next, and so on.

Of course, once your piece is accepted for publication, an editor will help you see where it can be stronger, and you'll need to tinker a bit more. But once that's over, you'll have something worth publishing: a tour in the literary universe worth taking.

TWO MASTER TOUR GUIDES

The beginnings of cosmic ray studies in Utah—the ancestor of the current Telescope Array Project—began in the 1960s at the University of Utah with Jack W. Keuffel and Haven Eldred Bergeson. In 1969, Keuffel and Bergeson won the Willard Gardner Prize and $2,000 for their contributions to cosmic ray studies, and they were each asked to give a speech to a broad academic audience. Keuffel and Bergeson wrote their speeches (which they knew would also be published and distributed) with sound, rhythm, and pacing in mind, with the end goal of explaining complicated particle physics to people outside their discipline.

In Keuffel's speech, after graciously accepting the prize and saying nice things about his adopted state of Utah, he talks about how important the Utah mountains and the silver mines underneath them had been for their cosmic ray experiment. "Why then mountains?" he says at the start of a new paragraph, and he launches into the details of his experiment. I've set repeated key words in boldface and underlined connecting words to show the trail of dots Keuffel leaves for the reader.

> Why then **mountains**? It is because **mountains** serve as an
> **energy** analyzer and also as a shield. The cosmic-ray beam
> contains particles of all **energies**, but those with the high
> **energies** of interest to us are infrequent compared to the
> low-**energy** background. By operating under a **mountain**,
> we shield our apparatus against the rain of low-**energy**
> cosmic rays, which would otherwise swamp our observa-
> tions. At the same time, the different thicknesses of the
> **mountain** enable us to study different **energies** of high-
> **energy** cosmic rays.

This is more of a "summary" rather than a "scene" that tells a full story, but it is a very clear explanation. Actors are acting: *mountains serve,* the *beam contains, we shield,* the *rain* that *would swamp.* The main words—*mountain* and *energy*—leave a connecting trail of key words for the reader to follow. By the end of the paragraph, the sentence "Why then mountains?" has been answered: the mountain lets in some energy from cosmic ray particles but not other energies. This is paragraph coherence in action; Keuffel leaves enough dots for us to connect.

Bergeson, in his speech, was similarly gracious, saying: "I feel a little bit like a fellow who might have been congratulated for building the pyramids. It was all done with slave labor." After thanking the "large number of extremely intelligent, dedicated, and ingenious graduate students," Bergeson then says this, which tells me he is fully aware of his primary audience: "It is generally thought that when scientists speak that only specialists in their field can understand them. I'd like to try to prove today that this is not necessarily so. I hope that my colleagues in the arts and letters will find what is said here today interesting and enlightening." Bergeson then proceeds to paint the story of cosmic rays and particle physics and, just as Keuffel did, Bergeson leaves enough dots for us to connect the various paragraphs together. Here is the first sentence of each of his subsequent paragraphs:

- The story of cosmic rays really begins shortly after the year 1900.
- Radioactivity provided a good explanation for the discharge of electrometers. However . . .
- One great skeptic was Robert A. Millikan.
- The next great surprise occurred in the early 1930s.
- Nature's nastiest cosmic ray trick came to light in the late 1930s.
- For the next cosmic ray story, we have to go back in time to 1931 and begin the story with research outside the field of cosmic rays.
- By the late 1950s, a very neat theory for neutrino interactions had been worked out.
- In the University of Utah neutrino detector in Park City, we have detected our first neutrino interactions this year. [This is a one-sentence paragraph, meant to bring emphasis to this one sentence.]

- In 1962, Professor Jack W. Keuffel had the ingenious idea that it would be possible to study two of nature's most mysterious particles—the muons and the neutrinos—simultaneously with one piece of apparatus located at a modest depth beneath the surface of the earth.
- Neutrinos have such a high range that most of them will pass completely through the earth without doing anything.
- When Keuffel first conceived this experiment, he had one very severe problem: how can you build a huge apparatus within the limits of the available money? To do so, he had to suggest two new inventions. One was . . .
- The second thing he suggested was a new kind of spark chamber.
- We also had to develop new data handling systems because no experiment had ever been performed in cosmic rays with the wealth and complexity of data that are obtained with this apparatus.
- The whole experiment has been very exciting, very surprising, a great deal of effort, and a lot of fun.

What we can see in these topic sentences alone is enough for us to get a broad outline of the whole story. Coherence and pacing also apply across paragraphs, not just within them: there are long paragraphs and short paragraphs, conjunctive adverbs as signposts, and a balance of summary and scene.

Bergeson ends his speech at the University of Utah—knowing many in his audience were likely religious and had a decent sense of humor—like this:

I'd like to conclude by reading a little something that I wrote expressing my feeling about all of this. If this were just an academy of science, I might be tempted to call it a poem. But there are those here who would know better, so I'll have to admit that what I've written is a little doggerel. It is based on another mystery of cosmic rays that I have not mentioned, and that is this: we don't know where they come from. They arrive at extremely high energies—fantastically high energies—from all directions in space. Now, scientists are never without an explanation. They may be wrong, but there is always an explanation. The current explanation is that cosmic rays come from a little object called a pulsar, which is the remnant of an explosion of a star. The pulsar is thought to be a rapidly spinning star composed mostly of neutrons. The verse is entitled "Cosmic Guile."

> Twinkle, twinkle small pulsar,
> Now I know just what you are.
> Quickly turning neutron star,
> Away in space, but not too far.
>
> As you turn you shed the bits
> Of matter that will try our wits,
> And lead us on until we're seeing
> The final nature of all being.
>
> Or is it a celestial hoax
> Planned by those great far-off Folks,
> A set-up job to make us stumble,
> To lift us up, then make us humble.
>
> No, we've known them for quite a while.
> Viciousness is not their style.
> When we err, they surely smile,
> Entice us on. They have no guile.

While I'm pretty sure Dyna's middle name is Cosmic Guile, I loved that Keuffel and Bergeson—pillars in particle physics at the University of

Utah, progenitors of the ultrahigh-energy cosmic ray experiment I now work for—knew how to form paragraphs and knew what their audience would appreciate.

THE FINAL STEP: CREATE A "GRAPHITE" PARAGRAPH

As we built from a **word**, to a **phrase**, to a **clause**, to a **sentence**, to a **super-sentence**, we likewise showed increasing complexity by building from a **quark**, to a **baryon** (a proton or neutron), to a fused **nucleus** (a proton plus a neutron), to a complete **deuterium** atom (a proton and neutron plus leptons), to a stabilized **carbon** atom (six protons, six neutrons, and six electrons).

To practice a model carbon sentence, abandon the strictures we've previously placed on exact words in exact phrases conforming to exact syntactical roles: they were scaffolds to help you learn the principle, and we can take away the scaffold now. Practice making a carbon sentence by writing a complete sentence composed of 12 words and 6 punctuation marks, scanned:

Science writers, with these six lessons,
will (surely!) make billions of dollars.

I didn't say the sentence had to be true.

You can go up to iron . . . or beyond . . . with longer sentences, if you're careful: iron has 26 protons and 30 neutrons, plus 26 electrons in four different orbital shells. It makes for good writing practice—a good word puzzle—to limit yourself to a certain number of words and punctuation marks in a sentence.

Ideally, in any paragraph we want to vary sentence length, but let's take carbon as our average sentence-atom. For our final step—combining sentences into a paragraph—let's return to Keith Houston's book *Shady Characters: The Secret Life of Punctuation, Symbols & Other Typographical Marks*, which we briefly discussed in Lesson IV.

Here is the origin story of the paragraph, as telegraphed by the now-invisible pilcrow mark (¶):

> The *paragraphos*, from the Greek *para-*, "beside," and *graphein*, "write," first appeared around the fourth century BC and took the form of a horizontal line or angle in the margin to the left of the main text. The exact meaning of the *paragraphos* varied with the context in which it was used and the proclivities of the author, but most often it marked a change of topic or structure: in drama it might denote a change of speaker, in poetry a new stanza, and in an everyday document it could demarcate anything from a new section to the end of a *periodos* [period].

To create a Subatomic Writing paragraph, let's take our atom of carbon and multiply it by, say, six. If we chain a string of carbon atoms together, we'll end up with six atoms in a ring, six model sentences in a well-paced, cohesive, and concise paragraph. The ring begins with an interesting or informative topic sentence, continues with other sentences that prove or expound on the topic sentence, and ends with a final sentence that simultaneously connects back to the topic sentence and can link forward to the next topic.

Congratulations! You've just created one layer of literary graphite.

Graphite is a crystalline element, meaning a solid substance of matter containing repeating units, and it forms sheets of carbon that are one atom thick. Ever wonder why the graphite used in pencils has a slippery feel? The hexagonal ring-sheets of carbon aren't connected tightly (unlike a diamond's atomic structure—also carbon—which is linked solidly into cubes). The loose connection between sheets allows the layers to slide off each other, leaving a gray mark on the paper. *Graphite*, like *paragraph*, comes from the Greek γράφειν (*gráphein*), "to draw, to write" (the same root as *grammar*, *graphs*, *diagrams*, and *graphemes*). English may be Germanic, but our processes of both science and writing have unquestionably been influenced by ancient Greek thought.

After you've completed one paragraph, go subatomic on another, linking it into another layer of graphite loosely connected to the first. Sort, fuse, and link again and again, unit by unit, until your finished piece hums with vibrational energy, stands strong in phrases glued together, creates an electric charge through its clauses, amplifies clauses with punctuation and other conventions, rolls off the tongue with repetition and variation, and finally, coalesces together in moving, cohesive paragraphs. In forming the substance of graphite—which is also a superb conductor of heat and electricity—you've just created the means whereby your reader can connect the dots and successfully see the form of your message: the heart of a pencil.

EXERCISE 1

Can you create a carbon sentence that teaches something about science? A carbon sentence in Subatomic Writing is 12 words and 6 punctuation marks. Or, if you'd really like a challenge, break it down to a quark count: a carbon sentence is composed of 18 Ups and 18 Downs—18 Lexical words and 18 Function words, for a total of 36 words—interspersed with six punctuation marks in the sentence. You can scatter capitalization, emphasis, and spaces as you need.

I'd love to see your carbon sentences, so if you figure out this puzzle, post your sentence-atom to Twitter with the tags #SubatomicWriting and @jamiezvirzdin.

EXERCISE 2

Can you create a graphite paragraph? Use your carbon sentence and add five more sentences (of varying word count) around it, for a paragraph consisting of six sentences that amplify, direct, expand, and connect with each other. Are some of your sentences short and some long? How have you managed the pacing along the way so the reader doesn't get bored? How are you logically connecting the sentences together? Have you placed a topic sentence at the beginning of the paragraph? Do you introduce known information at the beginning of the sentence and end with new information? Have you used transition words (like conjunctive adverbs)? Have you used too many? Have you left a trail of repeating key words, synonyms, and pronouns for the reader to follow? Can you cut down on any extra words?

EXERCISE 3

If you know another language, how is its literary universe different than the English universe? Do paragraphs flow differently in that universe?

EXERCISE 4

Look at the length of your favorite author's sentences and paragraphs. Find a page you particularly like and count the number of words in each sentence and the number of sentences in each paragraph. How are they connecting them together?

THE THEORY
OF EVERYTHING

Here let me pause.—These transient facts,
These fugitive impressions,
Must be transformed by mental acts,
To permanent possessions.
Then summon up your grasp of mind,
Your fancy scientific,
Till sights and sounds with thought combine
Become of truth prolific.

Go to! prepare your mental bricks,
Fetch them from every quarter,
Firm on the sand your basement fix
With best sensation mortar.
The top shall rise to heaven on high—
Or such an elevation,
That the swift whirl with which we fly
Shall conquer gravitation.

—James Clerk Maxwell, "To the Chief Musician upon Nabla: A Tyndallic Ode"

HALLOWEEN HAS ARRIVED, and I've finished the six lessons. They aren't perfect, but they are the best I can do.

I've done as the demon has asked—rather, demanded. I worked out a model—not a perfect model, but no model is—of *how* particles of the English language are like physical particles of matter. Maybe a publisher would like it. Maybe, down the line, a reader—maybe you, O brave one, you who have survived the vicissitudes of building both grammar and graphite in these pages—will use Subatomic Writing to reduce so much communicative entropy in the world that it would astound Claude Shannon, the mathematician and engineer who developed the concept of information entropy.

I've also fulfilled my parental duty to walk in crisp autumn air and leaves dressed as a Sesame Street Martian with Max. Andrew, not a fan of Halloween, read in the basement with the lights off. Back home—loot divided between the Martians—the trickle of Halloweeners stopped, and Andrew read the *Iliad* to Max before they both fell asleep.

Now I am waiting in the library for Dyna, printed manuscript in hand.

In fact, I've been waiting for three hours for her to show up, and it's almost midnight. "The muses are ghosts, and sometimes they come uninvited," Stephen King said in *Bag of Bones*, but tonight I want my lazy, good-for-nothing muse to appear so we can finish this speculative thought experiment.

While I'm waiting here on the beanbag, I might as well scribble some final thoughts about Quantum Field Theory and the Theory of Everything. We've gone so subatomic that, like Marvel's Ant-Man, we might get lost forever unless we zoom back out a bit.

Quantum Field Theory (QFT) is an important-sounding name, but all it means is that any two particles, or systems of particles, are connected by interactive fields, those bosons we talked about. Fields, by the way, is itself a metaphor, coined in 1845 by Michael Faraday, he who was willing to "leave the strict line of reasoning for a time, and enter upon a few speculations." I'm glad he did, but I'm also grateful for the others, like James Clerk Maxwell, who came after and verified Faraday's speculations.

While we can successfully use literary gravitons to shape our sentences into coherent sheets of paragraphs, the trick now, in the physical universe, is to connect QFT to General Relativity, which describes how gravity works. We hope, one day, to incorporate all four forces (if there are indeed only four—I'm always suspicious there might be something else lurking out there) into a unified **Theory of Everything**. It was this "superforce" that governed the universe in the first instant after the Big Bang, after which it split apart into the four forces we think we understand.

It's a lofty goal, along with all the other unsolved mysteries in science. Is it possible? I think so, if we stick to the scientific method as best as we can and communicate better and more effectively with each other. Somewhere on this globe there's another James Clerk Maxwell or John Dalton or Lise Meitner or Albert Einstein or Vera Rubin, and they not only think outside the box but also share—or have the potential to share—those unboxed thoughts coherently with others. Just as a paragraph can describe more than one word can, effective collaboration with others is how we can move forward in science.

In the meantime, in physics, we still have plenty of questions concerning the Standard Model we already have, like why Z and W bosons have mass but the photon doesn't, why W has more mass than we thought it did, and why the electromagnetic force and weak force are distinct at low energies but similar at high energies (combining into the *electroweak* force). We're still in the process of proving that the strong force and the electroweak force can be combined into a **Grand Unified Theory (GUT)**. Gravity, however, remains the missing piece of the puzzle. If there are four puzzle pieces and one is missing, it starts to grate on your nerves after a while.

As we continue to discover the nature of our existence, I hope you spare a passing thought for how the literary universe interacts with the physical universe. Quantum things aren't balls, or lines, or words, or whatever common-word analogy we apply to the quantum world, but models bring us closer to understanding. If you find a better model, write about it and share it.

I've focused almost exclusively this week on the quantum elements of writing, but if you're interested in "classical" science writing, the Knight Science Journalism (KSJ) program at MIT has a free book on science writing issues, the *KSJ Science Editing Handbook*. You can access it here: https://ksjhandbook.org/. It can assist you in "macro" issues of science writing, like helping readers understand statistics, for example, or how to best address controversies, how to find a narrative structure without resorting to demons, and much more.

I'm all for learning how to write better, but in the end, you must sit down for extended periods of time—or stand, if you're lucky enough to have a stand-up desk—and write bad drafts. Then, with the principles of Subatomic Writing, make those drafts a little better. When you feel overwhelmed, remember to come back to motion, the force at the ancestral heart of writing. I find it helps to read your work out loud pretending to be someone else you know, someone who might move in a different frame of reference than you do. How might your words move along for them?

However, know that you will never please everyone. Give your writing to one friend and they love it; give it to the next and they think it's garbage. So, yes, do think of your reader, use language you believe will matter to them; however, if you find you're becoming paralyzed by always trying to fit into everyone's frame of reference, remember: you can't. There is a surprising amount of peace and happiness in a healthy dose of cosmic fatalism.

Maybe that's the most important point I can pass along: while it's good to give your draft a subatomic sweep and work toward being the best writer you can be, it doesn't need to be perfect. Just turn it in. If a publisher says no, try again somewhere else. If lots of publishers say no, revise, and then try again. The need to appear perfect and perfectly knowledgeable and perfectly grammatical (in whatever style), in front of colleagues or anyone else, has kept many a great physicist, science writer, and literary fiction, nonfiction, and poetry writer from moving forward and making a contribution to our world through writing.

Where *is* she? To keep my eyes open, I pulled from my library shelf a book containing facsimiles of letters Maxwell had written to his friend Tait. On December 11, 1867, Maxwell wrote, "Now conceive a finite being who knows the paths and velocities of all the molecules by simple inspection but who can do no work except open and close a hole in the diaphragm by means of a slide without mass."

This was the origin of Maxwell's thought experiment, born in a letter written to his friend. I turned the page and saw another letter to Tait from around the same time:

Catechism:

Concerning Demons.

1. Who gave them this name? Thomson.
2. What were they by nature? Very small BUT lively beings incapable of doing work but able to open and shut valves which move without friction or inertia.
3. What was their chief end? To show that the 2nd Law of Thermodynamics has only a statistical certainty.
4. Is the production of an inequality of temperature their only occupation? No, for less intelligent demons can produce a difference in pressure as well as temperature by merely allowing all particles going in one direction while stopping all those going the other way. This reduces the demon to a valve. As such value him. Call him no more a demon but a valve like that of the hydraulic ram, suppose.

"Pfft, a valve. How insulting. Glad I'm not *his* demon," I heard. Dyna was lounging in my chair. She seemed sober this time, cleaned up.

She was wearing white shorts and a navy-blue T-shirt that said, "Is not life a hundred times too short for us—to bore ourselves?—Nietzsche."

"There you are! Finally. It's after midnight!"

"Well, I had to make my final entrance at 12:34:56 am, you doink. You know, for mnemonic consistency. So, where's the manuscript? All done? Six lessons?"

"Here it is." I handed the manuscript to her. She wiggled the stack back and forth, ran a long nail across the first page, then threw the whole stack into the air, watching all my hard work flutter to the ground.

"Sounds OK," she said. "Turn it in." Leaning over to witness the mess she'd just created, she pointed to a page of Lesson III. "You did forget to cover the rest of the irrealis moods, particularly the conditionals. And I'm not seeing anything on all the types of cleft sentences. And over here, there's nothing about the Pauli exclusion principle and degenerate matter—my favorite type of matter. I guess we'll save that for next time."

Fuming, I bent to pick up the pages. "There will be no next—" I froze as I heard a creak on the stairs. The sound of creaking footsteps intensified, and before I could prevent it, there was Max, at the doorway of the library.

"Hey, Dyna," he said. "Sorry, I woke up a little late."

I stared at Max. "You know this creature?" I gestured toward the demon in my chair.

Dyna examined her nails with two eyes, while the third, half-lidded, strayed guiltily in my direction. "I might have felt a little bad Tuesday night, so I popped in and explained the situation to him. Told him when I'd be back. Fortunately, your kid's more of a sport than you are." She pointed to the highest level of the cat tree.

In a radiant shower of sparks, Tom appeared, rethroned. He was at least several pounds heavier than he was a week ago, but otherwise seemed no worse for wear.

Relief poured through me. "I see you fed him, at least," I said. The cat stretched, then jumped—oozed—down the cat tree, onto my chair, and into Dyna's lap.

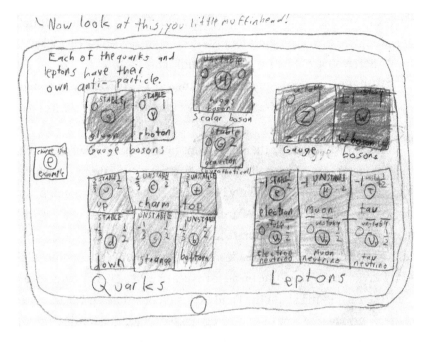

FIGURE 7.1 Max's sketch recounting Dyna's idea
of "teaching" the Standard Model.
Source: Maxwell Zvirzdin.

"Oh, well, I put a vat of raw meat in the sealed box, next to the flask of hydrocyanic acid, if that's what you mean."

Max came over to scratch Tom under the chin, and my tabby purred louder than a chainsaw.

"By the way," Dyna said, "I may or may not have taught Max the Standard Model after I explained where Tom was. Hope you don't mind. It's not my job, it's yours, but he was woefully undereducated."

"Thank you, Dyna," I said, smiling stiffly.

Max hefted the cat onto his shoulder. "Yeah, it was fascinating. She's a pretty good teacher." He turned to me hopefully. "When do I get my own demon?"

"When you're older, dear," Dyna said, before I could say anything. "Sadly, however, it won't be me. I'm Maxwell's mother's demon."

I glared at her with the might of a thousand boiling stars.

"Hey Max," Dyna said, with all three eyes still on me. "Have you ever heard of a Little Willie poem?"

"OUT!" I said.

After she left, Max helped me stamp out the sparks, and then I reached out to hold my cat.

EPILOGUE

As an undergraduate, in the fall of 2007, I was writing a paper for an English course about poetry, and I wanted an excuse to use the math of matrices as the "critical theory" lens for interpreting a poem. English majors wield Marxism, feminism, formalism, structuralism, new historicism, postcolonialism, and all the other -*isms* to wrest meaning from literature (just as anyone in STEM can wrest meaning from statistical data), so why not use math to interpret literature too?

While thinking about this, I happened upon a curious article in the journal *Modern Language Studies.* The article, written by David Porush in 1990, gave me the excuse I needed to write my silly paper. The article was titled, "Eudoxical Discourse: A Post-Postmodern Model for the Relations between Science and Literature." Skimming through it—it was full of nominalizations and jargon—I found enough substance to quote it as an argument. So I wrote that crazy paper on poetry using matrices; I enjoyed writing it, and I believe I got an *A*. I also believe my use of Eudoxical discourse didn't permanently alienate my primary audience, since years later the professor found me on Facebook, and we're still friends.

I'd forgotten all about that matrix paper and Eudoxical discourse until I was 95 percent through writing this book—which definitely took longer than a week. (In fact, I'd been teaching Subatomic Writing to graduate students at Johns Hopkins for four years by that point, and I'd been thinking about particles of language and matter for a few years before that.) Remembering the matrix paper, I refound Porush's article and was delighted to reread the following:

These scientific theories and speculations [quantum mechanics, cybernetics, neural networks, irrational numbers,
and chaos theory], I contend, lead to a renewed appreciation
for the epistemological power of literary discourse. It sug-
gests a new view of science where the role of metaphor in its
description of reality is acknowledged and metaphor as an
alternative method for uncovering truths is embraced. This
new, more literary, more postmodern discourse I call Eudoxical discourse. . . . On the other hand, quantum theory has
required a reevaluation of the fundamental assumptions of
science, including the applicability of logic, the particulate
nature of matter, and the neutrality of the act of measure-
ment and so has provided a case study in how attachments to
metaphor impede epistemological progress.

To me, this means that across any discipline, metaphor is powerful—as is acknowledging that metaphors can also lead us down wrong
paths and may need rejiggering once in a while. But the fact remains
that we couldn't do physics or literature without metaphor, so we might
as well embrace it as we teach both, or even use one as a metaphor to
teach the other. I don't think Eudoxical discourse is everyone's cup of
tea, but I find it interesting that I fell into it again without meaning to.

Dyna is . . . Dyna. In a parallel universe in which I've Jekyll-Hyded
myself, she is the Hyde to my Jekyll. She was actually born—in our current universe—during ninth-grade biology class, on a day I was so
bored with the science words slumping out of the teacher's mouth that
I imagined a little blue demon leaping around the room, dumping out
the garbage, setting the desk on fire, hanging from the ceiling, flailing
papers everywhere. As strange as it sounds, her sideshow chaos helped
me listen to the teacher better.

I love the ancient Greek *dialectic*—a method of discussing oppos-
ing ideas to get closer to truth—even more than I love the idea of the

daimonic, the supernatural presence meant to assist in intelligent thought experiments, but in Dyna I could have both. Those ancient Greeks had a good thing going as they talked back and forth with each other. They started something unique in both grammar and science— as we think of them today—that continues to affect the way we perceive words and worlds.

Dyna *did* teach the real-life Max the particles of the Standard Model. I'd asked Max if he had any ideas for the end of this book, and he went away and came back a while later with the drawing you see in Figure 7.1. He either looked up all the details on the internet himself or he'd been paying close attention to my dinnertime ramblings—in any case, the story engaged him, too. This is the power of fiction in both teaching and learning science, and we're never too old nor too young for it. *Dyna*, after all, means "power" in Greek.

My tabby cat's actual name is Utah. While I did grow up in Utah, we got the cat Utah from a shelter in Delaware while living in Maryland. Since I thought this was confusing, I changed Utah's name to Tom for the story. We also have a blue merle Australian Shepherd puppy, Zoe, who has creepy-beautiful, light-blue eyes. The cat and dog are more or less friends.

Andrew dislikes Halloween, Dyna, and fiction in general, but since there's no sense arguing over a matter of taste, he is still the love of my life, and we've successfully practiced a vibrant dialectic for fifteen years now. Since we take turns reading to Max every night, Andrew just finished reading the *Iliad* to Max, and they're about halfway through the *Odyssey*. (Fun fact: those two ancient Greek epic poems were written in dactylic hexameter, and they are *super* violent.)

I did read all three volumes of Gary Larson cartoons with Max— to offset Homer!—and I did actually have a series of sleep tests but not for sleep paralysis. I just wake up a lot during the night, sometimes in a panic, for no good reason. I am jealous of people, like Max and Andrew, who practically sleep on command, but the silver lining is that such wakefulness makes writing and cosmic-ray night shifts for the Telescope Array Project easier and enjoyable.

In the spirit of all John Dalton did for the elements of both the physical and literary cosmos, I'll end with words from his 1801 grammar book:

> As language is the medium through which instruction in every science, and knowledge in general, must chiefly be communicated, an attention to its principles is of primary importance in every system of education. . . . If [this book] should generally meet with the approbation of the public, I shall be glad to receive and attend to good-natured criticism, in order to render it more worthy of their acceptance.

Thanks for going subatomic with me.

Acknowledgments

I'm still shocked and delighted that my associate director of the MA in Science Writing program at Johns Hopkins University, Melissa Hendricks, trusted me enough to create the Subatomic Writing course in 2018. Students earning particles instead of points? It was more than a quantum leap of faith. I have likewise enjoyed the continued support and encouragement of Karen Houppert, program director of the MA in Writing, the MA in Science Writing, and the MA in Teaching Writing at JHU.

More shock and delight followed as Subatomic Writing met with student approbation and requests for a textbook. With my acquisitions editor at Johns Hopkins University Press, Tiffany Gasbarrini, I laughed so much and worked so hard. Thanks also to the entire JHUP team for your help in pulling this book together. Pilar Wyman (http://www .wymanindexing.com/) created the amazing index for this book (not an easy task), and I still cannot believe my luck—my editor, Susan Matheson (https://www.susan-matheson.com/), is not only a seasoned academic and nonfiction editor but also a systems engineer who worked in data communications at AT&T Bell Laboratories. She took the text to a far better place than I ever could on my own.

One of my oldest and dearest friends, Brent Elmer, and his talented colleague Michael Bulla created the illustration of Dyna based on my description and a photo of my library chair. I laughed so much as Dyna became a more substantive part of this book.

I owe so much to my English and editing teachers over the years: to Beverly Labrum, who taught me how to diagram sentences in first grade; to Judy Mac, who made AP English a class where we laughed hard and read deeply; to every English teacher between first and twelfth grades who made us search dictionaries, memorize vocabulary, write poetry, and fill out grammar worksheets; to professors at BYU, like Bill Eggington, Mel Thorne, Debbie Harrison, Colleen Whitley, Beth Hedengren, Matthew Wickman, Patrick Madden, and Marv Gardner, who not only taught me the minutiae of writing and grammar and *The Chicago Manual of Style* but also how to be ethical, inclusive, creative, thorough, and professional; to my graduate professors and classmates at Bennington College, who taught me to loosen up and have more fun—and that fiction, nonfiction, poetry, and science are not mutually exclusive disciplines.

Same thanks to all my STEM teachers and mentors over the years: to my dad, for passing me math problems in church, problems that sometimes involved thought experiments, like Kermit the Frog's velocity as he fell off a ladder; to my mom, who signed me up for STEM camps and Science Days and programming competitions and drove me to all of them; to Mr. Romanello, my favorite physics teacher, whose daily motto was "Life sucks and then you die"; to the BYU professors who properly taught me about human evolution, to their everlasting credit; to my professors and classmates in the JHU Applied Physics graduate program; and to all my University of Utah colleagues at the Telescope Array Project, for your patience, your humor, and your dedication to solving the mysteries of the universe.

Thanks also to all my brilliant students and you, reader, for reading this whole weird little book. I'm eternally grateful for friends like Alastair, Karis, Ann Marie, Matt, Isaac, April, Liza, Rasha, Danuta, Rachel, Jerome, Steve, Nicole, Chad, Joanne, John, Pierre, and many others who gave constructive feedback on this textbook.

To my husband, Andrew, thanks for cheerfully enduring my late-night bookmaking and cosmic-ray shenanigans and making sure Dyna didn't consume me entirely. To Max, for being the most engaging, supportive, honest, hilarious, brilliant eleven-year-old a parent could ever wish for. And finally, to Dyna: thank you for becoming more of a presence in my thought experiments than I ever wanted, starting in ninth-grade biology class. You keep me learning.

Ideas for Using Subatomic Writing in the Classroom

This book can serve either as a primary text for writing and science writing courses or as a supplementary text for science courses. *Subatomic Writing* may be especially useful for students who struggled with writing in high school, who want a refresher, or who need to write a thesis or dissertation. While this book is not an introduction to the English language, students who are not native English speakers can benefit from the succinct definitions of language terms and a broad representation of how various particles of English relate to each other.

At the beginning of the semester, I ask my students to identify their subatomic weaknesses and strengths in writing and to set specific goals for improvement. Students then focus their attention on those particular lessons. I check in with students about their writing goals throughout the semester, and sometimes students change their goals once they encounter the lessons.

I also assign students the task of analyzing their favorite authors according to the information presented in each lesson. What is the author doing at the word level? What are they doing with word class and phrases? What are they doing with clauses, word order, punctuation, rhythm, emphasis, pacing, coherence? Students diagram and scan the sentence, and we talk through it together. Students then analyze their own sentences in the same way. The level of insight from these analyses continues to surprise and amaze me.

Professors can assign any of the exercises as warm-ups or homework. Writers may enjoy posting Subatomic Writing diagrams or poetic scansions of their favorite sentences to Twitter or Instagram. Assigning your own class hashtag along with #SubatomicWriting allows students to see what others are posting and to connect with other Subatomic Writers throughout the world. Students can also work in groups to compare sentence diagrams, quiz each other on terms, create visual summary sheets, and verbally explain how they sorted the various particles of English into their syntactical slots. Teachers can ask students to reread the "Unhallowed Origins" chapter and discuss what techniques were used to teach subatomic particle physics to the reader. Which science writing techniques did they find more successful? Less successful? What are best practices for communicating information in writing? What is the place of fiction and metaphor in science communication?

I encourage all readers of *Subatomic Writing* to write the same 700-word paper I assign my students. This paper, which should ideally touch on and teach science in some fashion, must be exactly 700 words long, not including the headings, title, notes, or bibliography. This limitation provides an incredible opportunity for writers to practice concision and all six lessons consciously: to delete or switch out boring words for more vibrant Lexical words (with Germanic etymologies to balance the Latin and Greek science jargon); to cogently nest words within phrases within clauses within sentences within paragraphs; to identify what types of valency patterns and clause combinations they gravitate toward; to not only punctuate so the reader's mind-lungs can breathe but also avoid puncturing the reader with too much punctuation; to infuse rhythm and emphasis in the sentences they want the reader to notice and remember; and to connect sentences logically and to vary the types and lengths of sentences to affect the pacing of a paragraph. The first draft(s) should be messy and free; thereafter, go subatomic on them, one lesson at a time. All writers already own a voice but, like a professional singer, you can exercise your voice to perform with greater range on the page.

While most of my students choose to write nonfiction essays on some topic of science, others choose poetry or even science fiction, and I happily allow this. I tell them that whatever science topic and genre they choose is fine, but I grade these final papers rigorously, giving them the full editorial experience. I encourage them to revise their papers according to my feedback and then submit the piece immediately to a publisher like *Brevity: A Journal of Concise Literary Nonfiction*. *Brevity* has also created a superb list of links to other flash nonfiction publications (https://brevitymag.com/where-to-publish-flash-nonfiction/).

The style guide we use in this book, *The Chicago Manual of Style*, 17th edition, is certainly not the only great style guide available to writers, but it is the current publishing standard at most publishing houses, including places that publish science articles and books. Because *Chicago* is so comprehensive and widely used, it offers a solid foundation for those who are seeking to publish any kind of writing. However, if you want your students to publish in a different style, googling "Difference between Chicago style and _____" can quickly show you the key differences.

Last but not least, I recommend that readers use any kind of multimedia to represent the demon named Dyna, to visualize her in a way—with symbols, scenes, objects, particles, letters, words, colors, spaces, punctuation—that memorably reinforces the information your students learn in the lessons. If they'd like to post their creations on social media, I assure you that Maxwell's mother's demon is not easily offended and loves the attention. You'd be doing me a major favor—such a distraction would finally allow me to get a decent night's sleep.

Appendix

Subatomic Writing Diagrams

The Horror That Is "That"

Consider the following grammatically correct sentence:

*The hag (1) **that** had nabbed my cat added (2) **that** it's not (3) **that** bad and (4) **that** (5) **that's** the only way (6) **that** I'll get (7) **that** cat back.*

In order of appearance, the word classes of *that* in this terrible sentence are

(1) a **pronoun** (a relativizer) used to kick off a relative clause: *[that had nabbed my cat]*. The antecedent for the relativizer (the noun the pronoun refers back to) is *hag*.

(2) a **complementizer** used purely as a Function word to introduce a complement clause functioning as a direct object: *added [**that** [it's not that bad]]*.

(3) an **adverb** modifying an adjective: *[that bad]*.

(4) another **complementizer** in parallel structure with the complementizer in sense (2): *added . . . [**that** [that's the only way . . .]]*.

(5) a **demonstrative pronoun** used as the subject of a complement clause: *[**that's** the only way . . .]*. This is an example of an *unclear antecedent*, meaning that the noun the pronoun refers back to is unclear.

(6) a **complementizer** introducing a complement clause that acts as an appositive to a noun, renaming the noun *way*: *the only way [**that** I'll get that cat back]*, and

(7) a **demonstrative determiner** (a boring adjective) modifying a noun: *[**that** cat]*.

Only (2) and (4) are used in the same way. Figure A.1 shows the Subatomic Writing diagram of this sentence. Tables 1.1, 2.2, and 3.1 list the abbreviations used in Subatomic Writing diagrams.

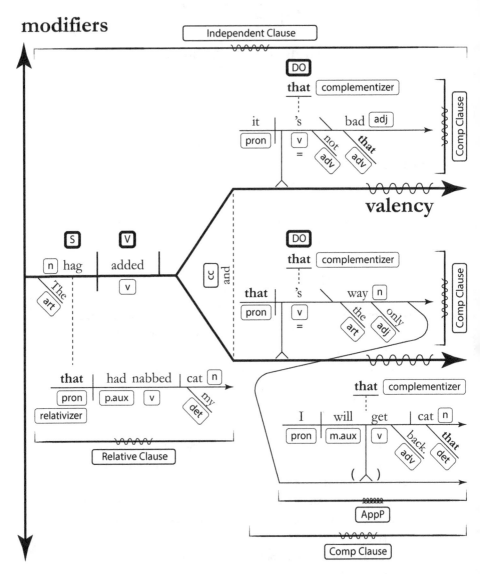

FIGURE A.1 A horrible sentence unraveled in a Subatomic Writing diagram and labeled by word class, clause, and valency pattern. *The hag that had nabbed my cat added that it's not that bad and that that's the only way that I will get that cat back.*

Source: Jamie Zvirzdin.

Whenever you can, avoid too many *thats* (no matter how they're being used) by substituting, deleting, or rewriting. When talking about a person—or a person-ish entity—use *who* instead of *that*, as in (1). Sometimes *that* can simply be dropped from the sentence but still be implied: we can safely drop (2), (4), and (6). (Sometimes the absence of *that* causes more confusion for the reader, so be careful.) We can change the pronoun *that* to *it* or use a stronger noun in (5). We can change *that* to another determiner like *the* or *my* in (7). Revised, we have: *The hag who nabbed my cat said it's not **that** bad and it's the only way I'll get my cat back.* It's still not a good sentence, but it won't make the reader's eye twitch. Finally, revised for clarity, concision, simplicity, and sound, a better iteration might be: *The hag who'd taken my cat said, "Writing a book is not that bad, and it's the only way you'll get your cat back."*

Diagramming Valency Patterns and Modifiers

Subatomic Writing (SW) diagrams continue the evolution of visualizing sentence geometry on a two-dimensional plane, an educational art form started by Clark, Reed, Kellogg, and many others. Small changes to previous models make diagramming easier and incorporate useful terminology adapted from *The Chicago Manual of Style*, 17th edition, as well as modern computational linguistics, most especially the *Longman Grammar of Spoken and Written English* by Douglas Biber, Susan Conrad, and Geoffrey Leech. However, sometimes SW diagrams diverge from naming conventions for simplicity's sake and to reduce confusion about similar-sounding terms.

SW diagrams also combine visual elements from Feynman diagrams, as described in Lesson III, to help students differentiate word classes, phrases, and clauses. Phrase labels (NP, VP, etc.) are marked with a spiral (the symbol for gluons from Feynman diagrams). Clauses (independent, dependent, complement, relative, etc.) are tagged with a wave-like symbol (the wave symbol represents photons in Feynman diagrams).

Students can tag sentences with abbreviations found in Tables 1.1, 2.2, and 3.1. Diagrams retain a sentence's capitalization and punctuation to show how punctuation interacts with phrases and clauses—and where there might be problems. *Chicago*-style punctuation of sentences are listed in Table 4.1.

To create an SW diagram, start with one primary horizontal line, called the **valency axis**, as shown in Figure A.2. The line perpendicular to the valency axis is the **modifier axis**. Since many sentences begin with introductory modifiers—not the independent clause's actual Subject and Verb—don't be fooled into thinking that a modifier deserves to be on the independent clause's valency axis.

The valency axis is where you build the sentence's first independent clause. The independent clause must have at least one Subject (S) and at least one Verb (V). The S + V will take up two slots on the main valency axis, divided by a bisecting, vertical line to create "slots" for those key syntactic roles. Other types of lines mark the connection between other syntactic roles.

You can create up to four slots (usually two or three) on one valency axis. The number and type of slots in the clause create its **valency pattern** (see Table 3.1 and Figure A.2). In a sentence diagram, display the valency pattern in order, even if it's not always in order in the actual sentence. (For example, questions have an inverted valency pattern. To diagram a question, you'd first rephrase it as a statement. Relative clauses can likewise contain inversions.) Other types of clauses less important than the independent clause (dependent clause, relative clause, complement clause, etc., described in Lesson III) will have their own internal valency patterns on their own clause line above or below the core valency axis.

Take your sentence, find the head words of each syntactic role, and identify the valency pattern. Then slide either one word, a few words (a phrase), or an entire clause (with its own nested valency pattern) into each slot. Think of the words in these valency slots as the **bare minimum** needed to communicate, something Tarzan might say to Jane: *Tarzan swings. Monkeys love bananas. Tarzan can give Jane kiss?*

Ask questions of the sentence to see which words should go in which slots. These questions have been listed in Lesson II. If you have five slots on any valency axis, you know that something is not right. At least one of them is a modifier or a nested clause and needs to be moved elsewhere.

Next, identify modifiers, which have a vertical component along the modifier axis. Diagram these modifiers under the word they modify, on a slanted line, drawn from top to bottom (a negative slope), starting underneath the valency axis. Draw words that modify nouns ("adjectivals," meaning words that function like adjectives) under nouns, and draw words that modify verbs ("adverbials," meaning words that function like adverbs) under verbs. The types of modifiers and how to recognize them are likewise listed in Lesson II.

Finally, ask if there are modifiers or Verbal Phrases that describe other modifiers or other Verbal Phrases. Figure A.3 shows as many modifiers as I could fit into one diagram. Too many modifiers can obviously overwhelm the syntax pattern at the heart of your core valency axis, the independent clause.

One more tip: Make sure all your prepositions are followed by a noun or a noun-like phrase or clause. This is the Object of the Preposition (OP). A preposition plus its OP equals a Prepositional Phrase.

modifiers

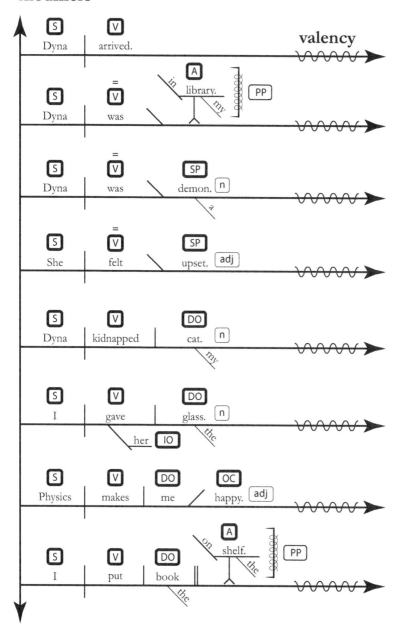

FIGURE A.2 A Subatomic Writing diagram showing types of possible valency patterns in clauses. See Table 3.1 for comparison. Notice the orientation and length of lines that divide the slots from each other. When my students use these distinctive lines in these distinctive orders, I know that they know what the valency pattern of each clause is.

Source: Jamie Zvirzdin.

modifiers

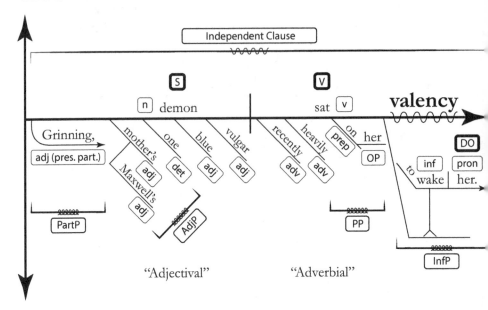

FIGURE A.3 Various types of modifiers in a Subatomic Writing diagram.
This sentence has a valency pattern of Subject + Verb (S + V), with a host of
(excessive) modifiers: *Grinning, Maxwell's mother's one blue vulgar demon
recently sat heavily on her to wake her.*
Source: Jamie Zvirzdin.

Diagramming Compounds versus Diagramming Multiple Independent Clauses

Any valency slot can split into more horizontal valency lines to accommodate parallel terms, creating a **compound** grouping of words, phrases, or clauses. Compound words in parallel slots are connected with a **dotted line** and a **coordinating
conjunction (cc)**. Sometimes the coordinating conjunction is *ellipsed*, or absent
from the sentence, but we still mark those absences with an *x*.

A **compound predicate**, for example, will look like S V cc V (Figure A.4). A
compound subject with five nouns, for example, will start with five horizontally
stacked lines at the beginning of the valency axis, all connected with a vertical dotted line and a coordinating conjunction where they start to merge. The stacked
lines then form one line, the valency axis; a vertical bisecting line comes next to
divide the noun from verb, and next comes the verb. Make sure the verb matches
each of the nouns (S, S, S, S, cc S V).

modifiers

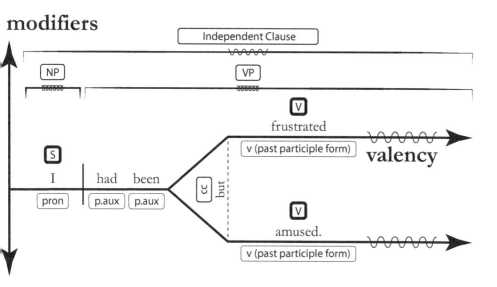

FIGURE A.4 An example of a compound predicate sentence:
I had been frustrated but amused. The Subject applies to both Verbs.
Notice that no comma comes after *frustrated.* While Figure A.4 was written
with passive-voice verbs (past tense, perfect aspect), Figure A.5 was
rewritten with active-voice verbs (still past tense, perfect aspect).
Source: Jamie Zvirzdin.

Any slot and any modifier can compound. Compound adjectives, compound
adverbs, compound Prepositional Phrases—these are all connected with a dotted
line and a coordinating conjunction. In Lesson III, I've provided links to websites
where you can study additional diagrams of all kinds.

While it's possible to have compound syntactic roles and compound modifiers, this is different than having a **compound sentence**, which is when the sentence has more than one independent clause. In compound sentences, two or
more valency axes are connected by a coordinating conjunction (cc) on a dotted
chair-like line that connects verb to verb (Figure A.5).

Recognizing the differences in sentence structure can not only improve your
writing, particularly punctuation, but also open the door to new ways to communicate science more effectively with readers.

Diagramming Dependent Clauses

If a clause does not fit into any slot along the core valency axis, it is often a dependent adverbial clause or a relative clause. Diagram these lesser clauses underneath
the independent clause's valency axis. Each clause will have its own mini-valency

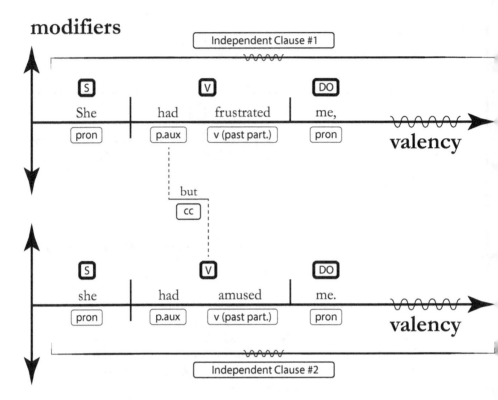

modifiers

FIGURE A.5 An example of a compound sentence: *She had frustrated me, but she had amused me.* This sentence contains two independent clauses, on their own valency axis, and they are linked by one of the seven coordinating conjunctions (FANBOYS: *for, and, nor, but, or, yet, so*). Unless the first independent clause is extremely short, we always put a comma before the coordinating conjunction (in this case, after the first direct object, *me*).

Source: Jamie Zvirzdin.

axis with its own valency pattern. In Figure A.1, a vertical dotted line connects the word *hag* to the pronoun renaming it in the relative clause—in this case, *that* (or *who*, if you consider the hag a person-like entity).

For dependent adverbial clauses, a dotted line links a verb from the independent clause to the verb in the dependent clause. Along the dotted line is a subordinating conjunction like *if, while, although, because, until,* or *whereas.* Some subordinating conjunctions double as prepositions or are used as complementizers, so be careful. Complement clauses often require their own nests, and their complementizers are placed on a line above the verb of the complement clause, connected with three dots. Lesson III provides other tips to help you identify the different kinds of dependent clauses.

The Verbals

Watch out for the **PIG: Participial Phrases**, **Infinitive Phrases**, and **Gerund Phrases**. They move all around the diagram depending on the specific sentence.

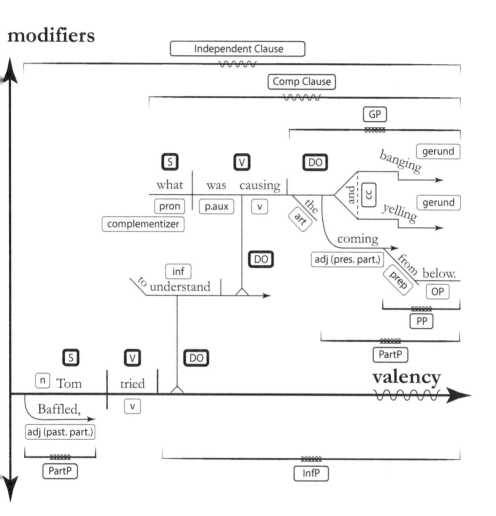

FIGURE A.6 An advanced Subatomic Writing diagram using all parts of the PIG plus a complement clause used as a direct object, headed by the *wh*-word complementizer *what*: *Baffled, Tom tried to understand what was causing the banging and yelling coming from below.* See Lesson II, "Verbals: The PIG" and "Verbal Phrases: The PIG Strikes Back" for a description of these monsters. See also Lesson III, "Dependent Clauses: The Tricky Stuff." *Noun Clause* is another name for a complement clause acting as a noun.

Source: Jamie Zvirzdin.

Ask: Is this PIG acting as a noun, verb, adjective, or adverbial? Is it one word, or does it appear in the sentence with its own nested valency pattern, like the adverbial Infinitive Phrase in Figure A.3? PIGs can have partial valency patterns, like zombified piglins from Minecraft (see Lesson II, "Verbals: The PIG" and "Verbals: The PIG Strikes Back"). As seen in Figure A.6, Verbal Phrases retain their verb-like power but can slide into a variety of slots, acting as syntactic roles or as modifiers.

Participial adjectives are diagrammed on a curved line that looks a bit like a **waterslide**. If the participle takes a direct object, separate the two with the same half line you see between valency axis verbs and direct objects, and place the noun-like direct object after the dividing line.

Infinitive Phrases are diagrammed almost like a Prepositional Phrase, like the **log ride** at an amusement park. Infinitives usually sit on stands, called **stems**, in the slot that they fill.

Gerund Phrases are diagrammed on a downward step like a little **waterfall** or **a very painful slide**.

If a complement clause slides into a slot, you'll see a stem that lifts the clause above its spot on the valency axis to give the clause more room. One of my students called this construction a **music stand**, which I rather like.

Miscellaneous

How do you do SW diagrams of direct address, conjunctive adverbs, dummy *there*, extraposed subjects, correlative conjunctions, elliptical adverb clauses . . . ? Technical terms never cease to multiply beyond what we have time and space to cover, and it's no different with grammar jargon and how to diagram every single particle.

Eugene Moutoux (http://www.german-latin-english.com/diagrams.htm) and others cover the minutiae of the minutiae if you get this far. Googling "direct address" and "sentence diagram" together—using quotation marks—will lead you to an image where you'll see the name of the direct address floating above the subject of the independent clause. Conjunctive adverbs and the dummy *there* also hang out in the upper-left corner of the diagram. Whatever gap lingers in your understanding—whatever information I couldn't cover satisfactorily in this textbook—seek to fill it with a quick search.

Welcome to the wonderful, maddening world of diagramming! Treat it like a puzzle, analyzing sentences you love and seeing how your own sentences measure up. Use diagramming to level up your power to communicate with others. Thanks for playing.

Selected Bibliography

Given the expansive nature of Subatomic Writing, I list here only the primary sources—of many—that I used to inform this textbook, writings I believe will be of most interest to those who wish to pursue the study of science and writing. For convenience, I've cited the location of direct quotes at the end of the bibliographic entry. See *The Chicago Manual of Style*, 17th edition, Section 14.64, for more about the various kinds of bibliographies and how to cite sources.

Unhallowed Origins

Ball, Philip [yes, Ball!]. "A New System of Chemical Philosophy." *Nature* 537 (September 2016): 32–33. https://doi.org/10.1038/537032a.

CERN. "The Standard Model." https://home.web.cern.ch/science/physics/standard -model.

Epstein, Lewis Carroll. *Thinking Physics: Understandable Practical Reality*. 3rd ed. San Francisco: Insight Press, 2015; quote on p. 1.

Falconer, Isobel. "Editing Cavendish: Maxwell and The Electrical Researches of Henry Cavendish." In *Proceedings of the First International Conference on the History of Physics—Trinity College Cambridge*, edited by M. J. Cooper, E. A. Davis, P. M. Schuster, and D. L. Weaire, April 2015. https://doi.org/10.48550 /arXiv.1504.07437.

Faraday, Michael. *Experimental Researches in Electricity.* In *Literature and Science in the Nineteenth Century: An Anthology*, edited by Laura Otis, 55–59. New York: Oxford University Press, 2009; quote on p. 56.

Gell-Mann, Murray. *The Quark and the Jaguar: Adventures in the Simple and the Complex*. New York: Henry Holt, 1995; quote on p. 180.

Houston, Keith. "The Mysterious Origins of Punctuation." *BBC Literature*, September 2, 2015. https://www.bbc.com/culture/article/20150902-the-mysterious -origins-of-punctuation.

Kundera, Milan. *The Book of Laughter and Forgetting*. New York: Harper Perennial, 1996.

Lovelace, Ada. *Sketch of the Analytical Engine*. In *Literature and Science in the Nineteenth Century: An Anthology*, edited by Laura Otis, 15–19. New York: Oxford University Press, 2009.

Maxwell, James Clerk. *Theory of Heat*. In *Literature and Science in the Nineteenth Century: An Anthology*, edited by Laura Otis, 70–73. New York: Oxford University Press, 2009.

Schrödinger, Erwin. *What Is Life? With Mind and Matter and Autobiographical Sketches*. With Contributions by Roger Penrose. London: Cambridge University Press, 1992; quote on p. 176.

Still, Ben. *Particle Physics: Brick by Brick*. Buffalo, NY: Firefly Books, 2018; quote on p. 128.

Tatter, Grace. "Playing to Learn: How a Pedagogy of Play Can Enliven the Classroom, for Students of All Ages." *Research Stories* (blog), Harvard University, March 11, 2019. https://www.gse.harvard.edu/news/uk/19/03/playing-learn; quote by Krechevsky.

Thomson, William (Lord Kelvin). *The Sorting Demon of Maxwell*. In *Literature and Science in the Nineteenth Century: An Anthology*, edited by Laura Otis, 79–81. New York: Oxford University Press, 2009; quote on pp. 80–81.

van der Horst, Pieter W. "The Omen of Sneezing." *Ancient Society* 43 (2013): 213–21. http://www.jstor.org/stable/44079975.

Lesson I. Good Vibrations

Biber, Douglas, Susan Conrad, and Geoffrey Leech. *Longman Grammar of Spoken and Written English*. New York: Pearson Education Limited, 2002.

Carroll, Sean. *The Particle at the End of the Universe: How the Hunt for the Higgs Boson Leads Us to the Edge of a New World*. New York: Penguin Group, 2012; quote on p. 38.

CERN. "The Brout-Englert-Higgs mechanism." https://home.cern/science/physics /higgs-boson.

The Chicago Manual of Style. 17th ed. Chicago: University of Chicago Press, 2017.

Churchill, Winston. "An Elder Statesman as Man of Letters." *New York Times Magazine*, November 13, 1949; quote on pp. 78–79. https://www.nytimes.com /1949/11/13/archives/an-elder-statesman-as-man-of-letters-mr-churchill-who-at -25-had.html.

Cuskley, Christine, Julia Simner, and Simon Kirby. "Phonological and Orthographic Influences in the *Bouba–Kiki* Effect." *Psychological Research* 81 (2017): 119–30. https://doi.org/10.1007/s00426-015-0709-2.

Ćwiek, Aleksandra, Susanne Fuchs, Christoph Draxler, et al. "The *Bouba/Kiki* Effect Is Robust across Cultures and Writing Systems." *Philosophical Transactions of the Royal Society B* 377, no. 1841 (January 2022). https://royalsocietypublishing .org/doi/10.1098/rstb.2020.0390.

Eakin, Emily. "Before the Word, Perhaps the Wink? Some Language Experts Think Humans Spoke First with Gestures." *New York Times*, May 18, 2002. https:// www.nytimes.com/2002/05/18/books/before-word-perhaps-wink-some -language-experts-think-humans-spoke-first-with.html.

Engelking, Carl. "In the Brain, Silent Reading Is the Same as Talking to Yourself." *Discover Magazine*, January 26, 2015. https://www.discovermagazine.com /mind/in-the-brain-silent-reading-is-the-same-as-talking-to-yourself.

Hasan, Ruqaiya. *Linguistics, Language, and Verbal Art.* Oxford: Oxford University Press, 1989.

Jones, Joe F., III. "Intelligible Matter and Geometry in Aristotle." *Apeiron* 17, no. 2 (1983): 94–102. https://doi.org/10.1515/APEIRON.1983.17.2.94.

Katzner, Kenneth. *The Languages of the World.* London: Billing & Sons, 1979.

Lieberman, Philip, and Robert McCarthy. "Tracking the Evolution of Language and Speech: Comparing Vocal Tracts to Identify Speech Capabilities." *Penn Museum Expedition Magazine* 49, no. 2 (2007): 15–20. https://www.penn .museum/documents/publications/expedition/PDFs/49-2/Lieberman.pdf.

Love, Shayla. "Why Are Letters Shaped the Way They Are?" *Vice*, February 11, 2022. https://www.vice.com/en/article/4awqz3/why-are-letters-shaped-the-way -they-are.

Maggio, Rosalie. *How They Said It: Wise and Witty Letters from the Famous and Infamous.* Hoboken, NJ: Prentice Hall Press, 1999; quote on p. 20.

Markel, Howard, interview by Ira Flatow. "Science Diction: The Origin of the Word 'Atom.'" *NPR*, November 19, 2010. https://www.npr.org/transcripts /131447080.

"The Mint: Research Projects: How Letters Got Their Shape." The Mint (Minds and Traditions Research Group), Max Planck Institute (website). https://www.shh .mpg.de/94549/themintgroup.

Montgomery, Scott L. *The Chicago Guide to Communicating Science.* 2nd ed. Chicago: University of Chicago Press, 2017; quote on p. 74.

Naeye, Robert. "Could Prion Stars Reveal a Hidden Reality?" *New Scientist*, February 13, 2008. https://www.newscientist.com/article/mg19726431-700 -could-preon-stars-reveal-a-hidden-reality/.

Oxford English Dictionary. 1978, s.v. "demon, n."

Oxford English Dictionary Online. March 2022, s.v. "demon, n. (and adj.)."

Perlman, Marcus, Rick Dale, and Gary Lupyan. "Iconicity Can Ground the Creation of Vocal Symbols." *Royal Society Open Science* 2 (2015): 150152. https:// royalsocietypublishing.org/doi/epdf/10.1098/rsos.150152.

Schwarz, John H. "String Theory." *Symmetry Magazine*, May 1, 2007. https://www .symmetrymagazine.org/article/may-2007/explain-it-in-60-seconds-string -theory.

Sword, Helen. "Zombie Nouns." *New York Times*, July 23, 2012. https://opinionator .blogs.nytimes.com/2012/07/23/zombie-nouns/.

Sutherland, Stephani. "When We Read, We Recognize Words as Pictures and Hear Them Spoken Aloud." *Scientific American*, July 1, 2015. https://www .scientificamerican.com/article/when-we-read-we-recognize-words-as-pictures -and-hear-them-spoken-aloud/.

Williams, Joseph, and Joseph Bizup. *Style: Lessons in Clarity and Grace.* 12th ed. New York: Pearson, 2016; quote on p. 127.

Wilson, Dave, dir. *Saturday Night Live.* Season 18, episode 11, skit: "Deep Thoughts by Jack Handey: Mankind." Aired January 16, 1993, on NBC. https://youtu.be /EigfHLZHQkE.

Lesson II. Nested Classes

Baer, Drake. "The Unexpectedly Existential Roots of Adjective Order." *The Cut*, September 7, 2016. https://www.thecut.com/2016/09/the-unexpectedly -existential-roots-of-adjective-order.html.

Biber, Douglas, Susan Conrad, and Geoffrey Leech. *Longman Grammar of Spoken and Written English.* New York: Pearson Education Limited, 2002.

Chase, A. W. *Dr. Chase's Recipes; or, Information for Everybody: An Invaluable Collection of About Eight Hundred Practical Recipes, for Merchants, Grocers, Saloon-Keepers, Physicians, Druggists, Tanners, Shoe Makers, Harness Makers, Painters, Jewelers, Blacksmiths, Tinners, Gunsmiths, Farriers, Barbers, Bakers, Dyers, Renovators, Farmers, and Families Generally.* Ann Arbor: A. W. Chase, 1867; quote on p. 342.

Klarreich, Erica. "A Tenacious Explorer of Abstract Surfaces." *Quanta Magazine*, August 12, 2014; includes quote by Mirzakhani. https://www.quantamagazine .org/maryam-mirzakhani-is-first-woman-fields-medalist-20140812/.

Kundera, Milan. *The Book of Laughter and Forgetting.* New York: Harper Perennial, 1996.

Muller, Derek. "Your Mass Is NOT from the Higgs Boson." Veritasium. May 8, 2013. YouTube video, 6:49. https://youtu.be/Ztc6QPNUqls.

Schoolhouse Rock Lyrics. "Conjunction Junction." First aired in 1973. https://www.schoolhouserock.tv/Conjunction.html.

Stamper, Kory. *Word by Word: The Secret Life of Dictionaries.* New York: Vintage, 2017; quote on p. 28.

Still, Ben. *Particle Physics: Brick by Brick.* Buffalo, NY: Firefly Books, 2018.

Zyla, P. A. et al. (Particle Data Group). *Progress of Theoretical and Experimental Physics.* London: Oxford University Press, 2020. https://pdg.lbl.gov.

Lesson III. Visual Syntax

Bennet, Jeffrey O., Megan O. Donahue, Nicolas Schneider, and Mark Voit. *The Cosmic Perspective.* 8th ed. New York: Pearson, 2016; quote on pp. 470–72.

Clark, Stephen. *A Practical Grammar: In Which Words, Phrases, and Sentences Are Classified according to Their Offices, and Their Relation to Each Other, Illustrated by a Complete System of Diagrams.* New York: A. S. Barnes, 1847; quote on p. iv.

Edwards, Tryon, ed. *A Dictionary of Thoughts: Being a Cyclopedia of Laconic Quotations from the Best Authors of the World, Both Ancient and Modern.* London: F. B. Dickerson, 1908; Southey quote on p. 52.

Fish, Stanley. *How to Write a Sentence: And How to Read One.* New York: Harper, 2011; Burgess quote on p. 2.

Florey, Kitty Burns. "A Picture of Language." *New York Times,* March 26, 2012. https://opinionator.blogs.nytimes.com/2012/03/26/a-picture-of-language/.

Gaither, Carl C., and Alma E. Cavazos-Gaither, eds. *Astronomically Speaking: A Dictionary of Quotations on Astronomy and Physics.* Bristol: Institute of Physics Publishing, 2003.

Kaiser, David. "Physics and Feynman's Diagrams." *American Scientist* 93 (March–April 2005); quote on p. 160. https://web.mit.edu/dikaiser/www/FdsAmSci.pdf.

Moutoux, Eugene R. "500 Sentence Diagrams: English Grammar and Usage." Personal website. http://www.german-latin-english.com/diagrams.htm.

Mulcahy, Rachel. "30 Quotes from Powerful Leaders to Celebrate Black History Month." *Southern Living,* n.d. https://www.southernliving.com/culture/black-history-quotes.

O'Brian, Bridget. "Prof. Stuart Firestein Explains Why Ignorance Is Central to Scientific Discovery." *Columbia News* (blog). Columbia University, May 8,

2012; includes Maxwell quote. https://news.columbia.edu/news/prof-stuart -firestein-explains-why-ignorance-central-scientific-discovery.

O'Brien, Elizabeth. "Diagramming Sentences Index." *Grammar Revolution* (website). https://www.english-grammar-revolution.com/diagramming -sentences.html.

Petit, Zachary. "21 Ray Bradbury Quotes: Your Moment of Friday Writing Zen." *Writer's Digest*, February 17, 2012.

Plait, Phil. "Crash Course Astronomy: Tripping the Light Fantastic." *Slate*, July 11, 2015. https://slate.com/technology/2015/07/crash-course-astronomy-light.html.

Schwab, I. R. "The Evolution of Eyes: Major Steps." *Eye* 32, no. 2 (2018): 302–313. https://www.nature.com/articles/eye2017226.

Walliman, Dominic. "How to Read Feynman Diagrams." Domain of Science. March 19, 2021. YouTube video, 9:18. https://youtu.be/oBNZOOuqO6c.

Lesson IV. Mind's Breath

Brown, Peter C., Henry L. Roediger, and Mark A. McDaniel. *Make It Stick: The Science of Successful Learning.* Cambridge, MA: Harvard University Press, 2014; quote on p. 207.

Dalton, John. *Elements of English Grammar, or A New System of Grammatical Instruction: For the Use of Schools and Academies.* London: R. & W. Dean, 1801; quote on p. 118. http://hdl.library.upenn.edu/1017/d/print/99214615 3503681.

———. *A New System of Chemical Philosophy.* Part I. Manchester: S. Russel, 1808. https://www.google.com/books/edition/A_New_System_of_Chemical _Philosophy/Wp7QAAAAMAAJ.

Houston, Keith. *Shady Characters: The Secret Life of Punctuation, Symbols & Other Typographical Marks.* New York: W. W. Norton, 2013.

Longknife, Ann, and K. D. Sullivan. *The Art of Styling Sentences.* 5th ed. Hauppauge, NY: Barron's Educational Series, 2012.

Lesson V. Repetition, Variation

Bryson, Bill. *A Short History of Nearly Everything.* New York: Broadway Books, 2005; quote on p. 7.

Crampton, Gertrude, ed. *Your Own Joke Book.* 16th ed. New York: Scholastic Book Services, 1972.

Illingworth, Sam. *A Sonnet to Science: Scientists and Their Poetry.* London: Manchester University Press, 2019; quotes on pp. 70–71, 77.

Khan, Salman. *The One World Schoolhouse: Education Reimagined*. London: Hodder & Stoughton, 2012; quote on p. 99.

Le Guin, Ursula K. *Steering the Craft: A Twenty-First-Century Guide to Sailing the Sea of Story*. New York: First Mariner Books, 2015; quote on p. 23.

Lobanov-Rostovsky, Sergei, ed. "The Poetics of Science." *Kenyon Review* 38, no. 5 (September–October 2016). https://kenyonreview.org/2016/09/poetics-of-science/.

"Making the Bible: The Press." *Treasures in Full Gutenberg Bible* (blog). The British Library, n.d. https://www.bl.uk/treasures/gutenberg/press.html.

Nave, Carl R. "Feynman Diagrams for Weak Force." *HyperPhysics*. Department of Physics and Astronomy, Georgia State University, 2016. http://hyperphysics .phy-astr.gsu.edu/hbase/Forces/funfor.html#c5.

Perrine, Laurence, and Thomas R. Arp. *Sound and Sense: An Introduction to Poetry*. 8th ed. Orlando: Harcourt Brace, 1992.

Serway, Raymond A., and John W. Jewett Jr. *Physics for Scientists and Engineers with Modern Physics*. 9th ed. Boston: Brooks/Cole Cengage Learning, 2014; quote on p. 1120.

Zyla, P. A. et al. (Particle Data Group). *Progress of Theoretical and Experimental Physics*. London: Oxford University Press, 2020; quote on p. 715. https://pdg .lbl.gov.

Lesson VI. Dot-to-Dot Game

Bergeson, H. E. "Gardner Prize Response: Probing Nature's Secrets with Cosmic Rays." In *Utah Academy Proceedings* 46, Part 2 (1969): 6–13.

Hewitt, Paul G. *Conceptual Physical Science*. 4th ed. New York: Pearson, 1981; includes Dirac quote on p. 51, footnote.

Houston, Keith. *Shady Characters: The Secret Life of Punctuation, Symbols & Other Typographical Marks*. New York: W. W. Norton, 2013; quote on p. 7.

Keuffel, J. W. "Gardner Prize Response: Cosmic Rays Under the Wasatch." In *Utah Academy Proceedings* 46, Part 2 (1969): 3–5.

King, Hobart M. "Graphite." *Geology.com*, n.d. https://geology.com/minerals /graphite.shtml.

Pinker, Steven. *The Sense of Style: The Thinking Person's Guide to Writing in the 21st Century*. New York: Penguin Books, 2014; quotes on pp. 139, 168; coherence relationships on pp. 159–66.

Serway, Raymond A., and John W. Jewett Jr. *Physics for Scientists and Engineers with Modern Physics*. 9th ed. Boston: Brooks/Cole Cengage Learning, 2014; quote on p. 1172.

The Theory of Everything

Blum, Deborah, Joshua Hatch, and Nicholas Jackson, eds. *The KSJ Science Editing Handbook*. Cambridge, MA: Knight Science Journalism (KSJ) Fellowship at MIT, 2022. https://ksjhandbook.org/.

Cep, Casey. "Science's Demons, from Descartes to Darwin and Beyond." *New Yorker*, January 8, 2021. https://www.newyorker.com/books/under-review /sciences-demons-from-descartes-to-darwin-and-beyond.

Griffiths, David. *Introduction to Elementary Particles*. 2nd ed. Hoboken, NJ: Wiley, 2014; quote on pp. 84–85.

Harman, P. M., ed. *The Scientific Letters and Papers of James Clerk Maxwell*. Volume 3, *1874–1879*. Cambridge: Cambridge University Press, 1990; quote on pp. 185–86; see also footnotes 1–4.

Maxwell, James Clerk. *Theory of Heat*. In *Literature and Science in the Nineteenth Century: An Anthology*, edited by Laura Otis, 70–73. New York: Oxford University Press, 2009; quote on p. 72.

———. "To the Chief Musician upon Nabla: A Tyndallic Ode." Poetry Foundation. https://www.poetryfoundation.org/poems/45778/to-the-chief-musician-upon -nabla-a-tyndallic-ode.

Nietzsche, Friedrich. *Beyond Good and Evil*. Project Gutenberg. https://www .gutenberg.org/files/4363/4363-h/4363-h.htm.

Epilogue

Dalton, John. *Elements of English Grammar, or A New System of Grammatical Instruction: For the Use of Schools and Academies*. London: R. & W. Dean, 1801; quote on p. xiii. http://hdl.library.upenn.edu/1017/d/print/992146153503681.

Porush, David. "Eudoxical Discourse: A Post-Postmodern Model for the Relations between Science and Literature." *Modern Language Studies* 20, no. 4 (Autumn 1990): 40–64; quote on p. 40. https://www.jstor.org/stable/3195060.

Index

Page numbers in **boldface** refer to figures and tables.